International Law and the Protection of "Climate Refugees"

"There are at least three reasons to read this book. The first is the topic. The protection of forced climate refugees is a growing problem which deserves attention. The book not only shows the strict interconnection among environmental protection and migration issues and the need to enhance international cooperation in these fields, but also how important it is to think out of schemes and in an interdisciplinary way. This is how the book approaches the topic. The second reason is the content. The book follows a path, which makes it clear, also to readers who are not familiar with environmental and migration international law, how to understand the problems related to the protection of climate migrants. It identifies those climate migrants who are in need of protection; it then explains why they are in this need and what kind of protection should be provided to them by considering the existing sources of international law granting, or capable of granting, such protection standards. The book then turns to the future and investigates the legal elements on which a forced climate migrant's law can (and hopefully will) be built. The third reason is the author: Giovanni Sciaccaluga is a profound and promising researcher, who very much cares about environmental protection and coherently lives his life. This book also mirrors the author's identity".

—Laura Carpaneto, *Associate Professor of European Union law, University of Genoa, Italy*

"While much has been written on the topic of climate change and displacement, the clear line of argument and strong interdisciplinary orientation, the up-to-date discussion of recent legal developments, and the insightful thoughts on how the law protecting people displaced across borders in the context of climate change might further evolve make Giovanni Sciaccaluga's book a particularly welcome contribution to the discussion".

—Walter Kälin, *Professor Emeritus, Institute of Public Law, University of Bern, Switzerland*

"This forward-looking work successfully sets the stage for the in-depth analysis of a central aspect of migration, one of the issues of most pressing concern to the international community today. By focusing on the relationship between International Law and Climate Change, the book greatly advances the contemporary discussion on human displacement across international frontiers and the protection under International Law of climate-change forced migrants, commonly referred to as 'climate refugees'. This canvas is filled by the separate extensive coverage given to migration management at the national level. The full picture that clearly emerges is solidly founded on thorough research in past and current migration patterns and the expanding corpus of applicable law, whether embodied in international or national hard- and soft-law instruments and judicial decisions. Their rigorous scientific study makes for a compelling argument in favor of the development in international law of a customary duty to protect the new category of 'climate refugees', to whom neither established refugee law nor protection mechanisms are currently applicable. By basing itself on the paramount concept of interdependence among States, the book epitomizes action most urgently needed to counter the forces that, by invoking an outmoded notion of the 'national interest', are bent on destroying rather than improving and, thus, strengthening the post war international legal order.

The comprehensive and systematic treatment of its crucial subject-matter, in an impeccable English prose, should incentivize the obligatory reading of this valuable book by the specialist—jurist as well as policy-maker—and the non specialist alike".

—Eduardo Valencia-Ospina, *Member and former Chairman of the International Law Commission and its Special Rapporteur on the "Protection of Persons in the Event of Disasters"*

"The interplay between climate change and migration is and will be a defining feature of our age. By unpacking this nexus, Sciaccaluga provides a compelling legal argument for the protection of forced climate migrants, which blends feasibility and ambition. A must read for any scholar, practitioner or layperson wishing to understand this complex relationship".

—Nathalie Tocci, *Director of the Istituto Affari Internazionali, Honorary Professor at the University of Tübingen, Germany, and Special Adviser to EU HRVP*

Giovanni Sciaccaluga

International Law and the Protection of "Climate Refugees"

palgrave
macmillan

Giovanni Sciaccaluga
University of Genoa
Genoa, Italy

ISBN 978-3-030-52401-2 ISBN 978-3-030-52402-9 (eBook)
https://doi.org/10.1007/978-3-030-52402-9

Cover credit: Maram/shutterstock.com

This Palgrave Macmillan imprint is published by the registered company Springer Nature Switzerland AG
The registered company address is: Gewerbestrasse 11, 6330 Cham, Switzerland

To my teachers

Foreword

Dulce Periculum: Reflections on Legal Evolution and Climate Migration at the End Times.

At the time of this writing a Global Pandemic rages. The pandemic in question: COVID-19—an insidious but ultimately, it seems, surmountable biological pathogen that has now infected some two million people globally and climbing. I, like most of the world, am presently in a government-imposed quarantine, engaging in what is now a global ritual of "social distancing" in an effort to slow the viruses spread and allow overwhelmed health services a chance to cope with patients.

To say that gives one time to reflect is an understatement. It also provides a unique perspective in history from which to think about crises more generally. That is to say, in the eye of the proverbial hurricane. In fact, our current dalliance with COVID-19 is but the latest crisis in a decade of crises, and *Zeitgeist* characterized by existential dread. The most recent, and most relevant to this study, is of course the refugee crisis. This has seen millions of people flee wars in Syria, Afghanistan, and Iraq for Europe and beyond at all costs. To what end? To reach the safety of a country that can protect them and their families from the atrocities of war and safe harbor from which new lives can be built.

In the aftermath of World War Two the resettlement of millions of refugees from countries that suffered hardest at the hands of the Nazis was not easy. European states were, as one might expect, preoccupied with their internal crises and chaos befalling their own citizens to take on others. Compassion fatigue, it seems, was a practical necessity. The international response to the refugee crisis was swift and took both legal and organizational forms. A United Nations Relief and Rehabilitation Administration (UNRRA) had been created in 1943 but succeeded by the International Refugee Organization in 1946; which in turn became the United Nations High Commission for Refugees in 1950.

Guaranteed in the Universal Declaration of Human Rights of 1948 was a "… right to seek and to enjoy in other countries asylum from persecution" and strict prohibitions on the arbitrary deprivations of nationality. However, with uncertainty as to what made a "refugee" different from a "migrant", the Geneva Convention on Refugees would end up defining the former as legal persons entitled to specific rights, including those that prohibited their *refoulement*—or forcible return—from countries of refuge. UNRRA was limited under its Articles of Agreement to assisting in the "repatriation or return" to their home countries of "displaced persons". It transported millions of former concentration-camp dwellers, forced laborers, and other victims of the Nazis to countries such as France, Belgium, and Greece.

Many of the Soviets departed willingly. Many had no choice. Their forcible return conflicted with the "*non-refoulement*" principle. Many "Displaced Persons"—or DPs—from Eastern Europe were taken over by Communists resisted repatriation. For those who did not find more immediate refuge in Western Europe, the United States, Canada, or Australia offered tantalizing options. The geopolitical complexities of The Cold War combined with predictions of increased labor demand, led Britain, Australia, and other countries to grant Poles and some other permanent settlement. The creation of the State of Israel in 1948 finally provided solace for Jews who had been driven from their homes in central and Eastern Europe. For them, the buoyant US economy held out the most tantalizing hope to refugees.

The more recent and undoubtedly more pernicious challenge of climate change helps us focus the crises of the past in the sobering light of the present. Be it natural disasters, rising sea levels, agricultural factors, automation, among others, these factors are driving people worldwide to leave their homes to survive. As is often the case, those hardest hit will likely be the poor whose labor is tied to the land. But there are always externalities when jobs start disappearing; crime starts paying when nothing else will, and eventually people get going before the going gets good.

The Map Is Not Territory

In his 1972 book, *Steps to an Ecology of Mind*, the British anthropologist and cybernetician Gregory Bateson developed his idea of a "difference that makes a difference" first laid out in a talk at Alfred Korzybski's Institute of General Semantics. His talk: "Form, Substance, and Difference" referred to Korzbyski's maxim "the map is not the territory". For Bateson "differences" were those key features that become parts of the "map". Practically, this meant that global effects were ultimately caused by the push and pull of physical forces. Yet in the formal world of patterns, and the cybernetic world of communications in particular:

> [W]hen you enter the world of communication, organization, etc., you leave behind that whole world in which effects are brought about by forces and impacts and energy exchange. You enter a world in which "effects"—and I am not sure one should still use the same word—are brought about by *differences*. That is, they are brought about by the sort of "thing" that gets onto the map from the territory. This is difference. (*Steps to an Ecology of Mind*, p. 452)

Yet the reason something becomes retrospectively significant in futures both near and far-flung is ultimately detached from the reason it was significant at the time of its creation—and it is almost always due to the fact that subsequent generations end up, and are entitled to, recalibrate what they accept as normative. Morality after all, it seems, extends no further than acceptability. But the challenges ahead can be informed by the past, at least in the hope that whatever choices we make going forward, be that at the level of government, law, society, or the individual, are inexorably bound up with the nature and quality of the future we create. Such is the tragedy of paradox.

Practically, this means the world will likely face hard choices. Some of these might be choices we have made before, for others; the past is no prologue. But if there are any general lessons to extract from history of political order and current accounts of its decay, it is that the autonomy of the legal institution is an essential condition for what Deakin calls the "impersonal and impartial application of the coercive power of the state, the 'rule of law' which is understood to be part of the bedrock of a market economy and democratic society". It is in these societies,

Deakin observes, that "the legal system informs and supports the web of beliefs and expectations through which societal coordination is combined with protection for the individual and private sphere". That this is so is, perhaps, a minor miracle. But it is precarious one, and one currently threatened by the politicization of legal decision-making, the potential collapse of the discursive sphere, and continued revisions to the neoliberal hypothesis about the economization of the legal system.

But this present crisis also makes another thing clear. It does not just matter what country you find yourself in during times of a crisis, what matters is what laws are in place to mitigate its worst effects, and whether the people see it as legitimate. But it is the legitimacy of social institutions that seems under most sustained threat. While there is no divining what lingering effects this pandemic might have, it seems safe to conclude it, like 9/11 before it, will be the basis for normalizing, and locking-in, various surveillance technologies to monitor global traffic flows, and put public health considerations at the top of the list for governments. What we might need then is a radical redistribution of resources from the surveillance-industrial complex to public health. But it would be a mistake to see public health as anything other than inseparable from the wider questions, contingences, and problems ahead.

While the history of law, and human experience itself, has largely been a creative process. At least in the sense we seem to be making it up as we go along. We must resist Napolean's fatalist assertion that history as nothing more than a "a set of lies agreed upon". Instead I urge the readers to consider Dr. Sciaccaluga's insights through the wisdom of Canadian media theorist Marshall McLuhan: "There is absolutely *no inevitability* as long as there is a willingness to contemplate what is happening". When we emerge out of this pandemic—and we *will* emerge from it—the world will be a different place, and the free movement of persons internationally will likely become a defining issue for international governments, NGOs, and indeed, all of us. It is thus perfect timing for Dr. Sciaccaluga's *International Law and the Protection of "Climate Refugees"* to cast sobering light on these most pressing of legal, societal, and existential challenges. I commend his work for its breadth of insight, analytical rigor, but mostly the essential humanity with which he addresses one of the most challenging issues of our times. We would all do well to heed not just his insights, but also the sincerity and conviction with which Dr. Sciaccaluga argues them. Just as COVID-19 pandemic has reminded us of our fragility as a species, the climate crisis should make clear one thing: we are *all* equal before a wave.

Summer 2020

Dr. Christopher Markou
Faculty of Law
University of Cambridge
Cambridge, UK

Acknowledgments

During the spring of 2014, I was wondering what topic to choose for my thesis. It was my last year of university. One day, I (un)fortunately lost a fingernail at home while closing a huge window in the living room, and, with my small injury, I went to the emergency room, where I patiently waited a couple of hours before seeing a doctor. While waiting, a TV monitor was looping the same show over and over again: it was "Home", an absorbing documentary on global environmental degradation. At some point, it said that the United Nations estimated that by 2050 there would be around 250 million "environmental refugees" worldwide. I had never heard of them before, but immediately got captured by this fascinating and ever-rising challenge. This was the exact moment I chose to address this topic. Six years and some research later, I am publishing my ideas on it.

I want to deeply thank Laura Carpaneto, who has patiently, wisely, and caringly accompanied me in this journey. My thanks also to Francesco Munari and Lorenzo Schiano di Pepe for their precious teachings in international environmental law, and to Pierangelo Celle for his in human rights law. I also want to thank all friends and colleagues who have made this book possible: Ilaria Queirolo, Francesco Pesce, Stefano Dominelli, and Francesca Maoli.

A thankful thought also goes to Jane McAdam, Walter Kälin, Sumudu Atapattu, and Benoit Mayer, whose work during the years has been of invaluable inspirational value.

To Nathalie Tocci for her tireless and decisive support in the creation of this book.

And to my family, for their emotional, cultural, and economic support throughout my entire life.

Contents

About the Author

Giovanni Sciaccaluga is a Ph.D. in social sciences and contract professor of International organizations law and of European Union law at the University of Genoa. He has published several articles on the topic of climate change-related migration. In 2015, he won the "International and comparative disaster law essay contest" of the International Federation of Red Cross and Red Crescent Societies with the paper "Climate change-related disasters and human displacement: towards an effective management system".

Abbreviations

ASEAN	Association of Southeast Asian Nations
CESCR	Committee on Economic, Social and Cultural Rights
COP	Conference of Parties
CRC	Convention on the Rights of the Child
EACH-FOR	Environmental Change & Forced Migration Scenarios
ECHR	European Convention on Human Rights
ECOWAS	Economic Community of West African States
EEM	Environmentally Emergency Migrant
EFM	Environmentally Forced Migrant
EMM	Environmentally Motivated Migrant
EPRS	European Parliamentary Research Services
EU	European Union
ExCom	Executive Committee
GCM	Global Compact for Safe, Orderly, and Regular Migration
GCR	Global Compact on Refugees
GHG	Greenhouse Gas
IASC	Inter-Agency Standing Committee
ICCPR	International Covenant on Civil and Political Rights
ICESCR	International Covenant on Economic, Social and Cultural Rights
ICJ	International Court of Justice
IDMC	Internal Displacement Monitoring Centre
IDP	Internally Displaced Person
IFRC	International Federation of Red Cross and Red Crescent Societies
ILC	International Law Commission

ILO	International Labour Organization
IOM	International Organization for Migration
IPCC	Intergovernmental Panel on Climate Change
NDC	Nationally Determined Contribution
NGO	Non-governmental organization
NRC	Norwegian Refugee Council
NZIPT	New Zealand Immigration and Protection Tribunal
OAU	Organization for African Unity
OECD	Organization for Economic Cooperation and Development
PAC	Pacific Access Category
PDD	Platform on Disaster Displacement
PPM	Part Per Million
RCP	Regional Consultative Process
RRTA	Refugee Review Tribunal of Australia
RSAA	Refugee Status Appeals Authority (New Zealand)
SADC	Southern African Development Community
TFD	Task Force on Displacement
UDHR	Universal Declaration of Human Rights
UN	United Nations
UNCBD	United Nations Convention on Biodiversity
UNCED	United Nations Conference on Environment and Development
UNDP	United Nations Development Programme
UN ECOSOC	United Nations Economic and Social Council
UNEP	United Nations Environment Programme
UNFCCC	United Nations Framework on Climate Change
UNGA	United Nations General Assembly
UNHCR	United Nations High Commissioner for Refugees
UNHRC	United Nations Human Rights Council
UNISDR	United Nations International Strategy for Disaster Reduction
UNOHCHR	United Nations Office of the High Commission on Human Rights
UNSC	United Nations Security Council
UNU	United Nations University
UNU-EHS	United Nations University—Institute of Environmental and Human Security
US	United States
WIM	Warsaw International Mechanism

CHAPTER 1

Introduction

It is clear today that climate change is having and will increasingly have an impact on human mobility, both at the internal and international level. According to the Internal Displacement Monitoring Center (IDMC) data, for instance, since 2009 a person per second has been displaced because of a natural disaster, with an average of 22.5 million people displaced by climate or weather-related events since 2008.[1]

Not surprisingly, these figures raise significant concerns about how the international community should deal with similar migration patterns. Moreover, it is worth recalling that the very first Intergovernmental Panel on Climate Change (IPCC) report, dating back to 1990, already highlighted that the single greatest consequence of global warming would precisely be on human mobility.

This book analyzes the topic of climate change-forced migrants (commonly referred to as "climate refugees") through the lens of public international law. The general aim is to understand the extent to which this branch of law can deal with one of the key global public policy questions of our age. Alas, climate science clearly shows that notwithstanding international mitigation efforts, global warming will keep increasing in

[1] For a detailed analysis of the numbers of persons involved in (disaster) displacement, see the IDMC website. Internal Displacement Monitoring Center, Global Internal Displacement Database. http://www.internal-displacement.org/database/displacement-data. Accessed 12 March 2020.

G. Sciaccaluga, *International Law and the Protection of "Climate Refugees"*,
https://doi.org/10.1007/978-3-030-52402-9_1

the next decades. As a side effect, more and more people will be either induced or forced to leave their homes or countries, and this poses significant challenges to the international community and its laws, which are stuck between the twofold obligation to protect inalienable human rights and the need to preserve state sovereignty and national identity.

Dedicated legal studies have proliferated over the last years, and research has clearly demonstrated that international law is currently not up to the task of protecting people forced to migrate due to climate change. This is basically because of the relative historical novelty and complexity of the latter: neither refugee law nor complementary protection mechanisms, both of which are essentially post-Holocaust legal frameworks, are currently applicable to "climate refugees" and this simply because climate and environmental degradations are not held as factors that can activate international protection mechanisms.

Truly speaking, the inapplicability of existing international protection systems also depends on the lack of a legally coherent conceptualization of the category of persons involved, who are commonly simplistically labeled as "climate refugees". The first objective of this book is thus to conceptualize this category in a manner that is relevant to international protection standards. As known, a legal definition of the category is for the time being still missing; on top, what remains unclear is how and why it should be considered as legally autonomous. This book builds on Sumudu Atapattu's definition of "forced climate migrants", understanding them as individuals compelled to leave their national state because its territory has (or is about to) become uninhabitable due to climate change-driven environmental degradations. By so doing, the book focuses on forced climate-related cross-border migrations that are characterized by a very limited chance of migrants returning to their countries of origin. In this sense, the 2018 Global Compact for Migration, while recognizing that adaptation is a priority, also points out that in some cases the possibility to return to the country of origin might not be possible for migrants, leading to the necessity to provide migration pathways or international protection standards for the affected individuals.

The consequent objective of this book is then to explain the reasons why "forced climate migrants" should be granted international protection. To substantiate this argument, the analysis focuses on climate change international law, and human rights law, pointing out the legal principles and rules upon which an international obligation to protect persons forced to migrate due to climate change is currently emerging under international law. Relying on instruments protecting human rights universally or regionally, this book advocates a legal obligation of states to protect forced climate migrants when origin countries have become extremely fragile (if not collapsed altogether) because of anthropogenic global warming. In this respect, for example, insular micro-states such as Kiribati or Tuvalu (sometimes referred to as "drowning" or "sinking" states) are the most affected ones. Science suggests that they will be "deterritorialized" by rising sea levels, generating a clear problem of governance of their populations, which will be compelled to leave their countries. Should these people not be protected by the international community and under international law if their state is to disappear because of "man-made" climate change?

Turning to the future, the book investigates the legal elements on which a "forced climate migrants' law" can be built. In this respect, it is possible to draw on the universal regulatory systems for the protection of human rights and on international climate change law. Considering these and referring also to the rules of international law in matters of responsibility, the book argues that within the international legal system a duty to provide some form of assistance to forced climate migrants in a third state is currently arising. The seeds of this are already in place: the progress made under the United Nations Framework Convention on Climate Change (UNFCCC), as well as intergovernmental initiatives such as the Nansen Initiative and the Platform on Disaster Displacement, and the adoption in 2018 of the Global Compacts for Refugees and Migrations go in this direction. At the same time, there is place for already existing legislation to be interpreted progressively by competent courts: some national cases and the possible evolution of the European Convention on Human Rights, for example, leave room for the reinterpretation of "forced climate migrants" as part of the larger group of individuals who can benefit from complementary protection frameworks. As some have pointed out, this is in fact very slow progress if we recall that the IPCC warned already thirty years ago that the greatest impact of climate change would be on human migration.[2] However, it is clear that the law, especially at the international level, somehow evolves "like a pachyderm"

[2] Atapattu (2018). A New Category of Refugees? "Climate Refugees" and Gaping Hole in International Law. In Behrman & Kent (Eds.), *"Climate refugees" Beyond the Legal Impasse?* New York: Routledge, p. 43.

because of its very nature, which always requires a minimum and common degree of coordination and agreement among states.

Our analysis starts with Chapter 2, which contextualizes the phenomenon of climate change-related migrations and illustrates the conceptual framework underpinning the book. After it, the work is composed of three different parts.

Part I reviews the most important findings concerning the link between climate change and migration to provide the reader with a sound analytical framework. The debate concerning climate-forced or induced migrations has been ongoing for decades now, and literature on the topic has proliferated over the years, also fueling a fair degree of confusion. In this sense, we are going to refer in this book to a symbolic "climate refugees cauldron": an intricate and confused set composed of all those who are directly or indirectly induced or forced to migrate because of climate change, be it within their national country or abroad. Part I aims at shedding some light in this cauldron, and proposes a conceptualization of the category of those who are forced to migrate internationally due to climate change, which is compatible with international protection systems. Chapter 3 draws attention to the most important findings concerning climate-related migration, investigating the etiologic link between climate change-driven environmental degradations and human mobility. Chapter 4 turns then to the different approaches adopted over the years to study the topic at hand. In fact, depending on the conceptualizations and definitions followed, radically different discourses and estimates have been proposed. Figures concerning future "climate refugees" span, for instance, from zero to hundreds of millions of people involved for the decades to come. The chapter discusses the pros and cons of the two main schools of thoughts that have studied the issue over time, maximalism and minimalism. This to shed further light in our "cauldron" and start drawing lines so as to appreciate that there actually are different categories of climate-related migrants, an important step to understand which ones can effectively be considered relevant through the prism of international protection law. Chapter 5 delves into a definitional analysis. The novelty of the phenomenon at stake has contributed to the proliferation of numerous (proto)legal classifications and definitions, such as "survival refugees", "climate refugees", "eco-migrants", "climate-induced migrants", or similar terms. By discussing and comparing these, the chapter proposes a definitional approach that might prove compatible with international protection standards.

Part II deals then with a quite pressing legal issue: why should individuals compelled to leave their countries because of climate change be protected abroad? One of the Gordian knots concerning climate-related migration arises from this simple but crucial question. This part of the book analyzes existing international law principles and rules to understand the extent to which forced climate migrants may be entitled to international protection. Its aim is to propose a comprehensive legal reasoning by virtue of which international law can, as was the case in past centuries with "political refugees", provide tools to protect a new category of forced migrants: the so-called "climate refugees". Chapter 6 argues that the evolution and gradual establishment of international climate change law leaves fewer and fewer opportunities to states to be indifferent to global warming. It consequently argues that the do no harm principle acquires growing importance when states consciously permit or promote on their territories climate-altering behaviors capable of damaging other countries. Building on this reasoning, Chapter 7 argues that different elements arising from international climate change law is bringing to light a new international custom, which obliges states to mitigate global warming and to adapt to it. Such obligations would moreover stand as *erga omnes* ones, and this because of the essential importance of the international "good" they aim at protecting: the climatic system in which the international community has developed to this day. If this theorization is valid, its impacts go far beyond the single issue of climate-related migrations. Focusing on this particular issue however, it is used to argue that under the umbrella of climate change adaptation policies and in fulfilment of existing human rights obligations, states should grant international protection to individuals compelled to leave their national territory because of climate change. Chapter 8 studies this topic further, highlighting the need for international action to avert the negative effects on human rights caused by climate change. Since the enjoyment of basic human rights such as life, health and access to livelihood can be infringed by climate-related environmental degradation, it is possible to link human rights law with the issue of forced climate migrants. Chapter 9 concludes Part II, arguing that in the (rare) cases in which severe climate-driven environmental alterations affect the entirety of a national territory, dismantling a state's capability to provide for its citizens, the international community should be obliged to protect these individuals through international protection frameworks. Basic human rights such as life, health and access to livelihoods are inalienable and non-derogable rights, not depending on the existence or recognition of any state. The main argument of this book is that in the name of inalienable human rights protection, forced climate migrants should be provided with international protection in third countries when their national state is environmentally weakened by climate change up to the point of becoming unable to grant the enjoyment of these rights to its citizens on its own territory.

Part III delves into the "embryos" of forced climate migrants' law. It looks at the legal frameworks under which a future protection and management system can grow in the following years. The progress made in this sense under the UNFCCC and the adoption in 2018 of the Global Compacts for Refugees and Migration witness that the international community is slowly equipping itself with instruments that address the issue of "climate refugees", both from the "protection" and "migration management" side, and Chapter 10 analyzes these evolutions. Starting from the 2010 Cancun Agreements, the UNFCCC has made some important steps in addressing climate-related displacement. The Chapter investigates these developments, which culminated in the endorsement of COP 24 in 2018 of the recommendations of the "Task Force on Displacement", addressing how they may fit in a systemic evolution of international law to provide responses to the challenges of forced climate migrations. The same is done with regard to the two Global Compacts adopted in 2018. These two pioneering instruments explicitly recognize for the first time in dedicated universal soft law tools that environmental pressures and climate change adverse effects are drivers of human migration and refugee movements. Chapter 11 finally analyzes complementary protection systems with the aim of understanding if they may adapt to protecting forced climate migrants in the years. Among existing international protection frameworks, complementary protection appears to be the fittest to this purpose. The chapter looks at relevant (inter)national case law to understand if competent courts have the room to progressively interpret current legislation on the prohibition of inhuman and degrading treatment to include in its application those who are forced to migrate internationally due to climate change.

The conclusion eventually sums up the thesis of the book. It highlights the work's most important features and research results, offering the reader a clear "stand-alone" final chapter.

CHAPTER 2

Conceptual Framework

Since the Eighties, the topic of migrants connected to climate change—generally known as "climate refugees"—has earned growing attention because of its political sensitivity, humanitarian significance, and high degree of complexity.[1] Over the years, literature on the topic has proliferated, with research coming to different and contrasting results with regard, for instance, to the number of people involved and even to the very nature of the phenomenon.[2] However, although the differences in

[1] The first recognition within the UN system of the phenomenon was by Essam El-Hinnawi. El-Hinnawi (1985). Environmental Refugees. Report. UNEP. https://digitallibrary.un.org/record/121267. Accessed 12 February 2020. In this regard, see Myers & Kent (1995) *Environmental Exodus: An Emergent Crisis in the Global Arena*. Washington, DC: Climate Institute. https://portals.iucn.org/library/node/22380. Accessed 17 March 2020.

[2] See, *ex multis*: McAdam (Ed.) (2010). *Climate Change and Displacement, Multidisciplinary Perspectives*. Oxford: Hart Publishing. See also Hugo (1995). Environmental Concerns and International Migration. *International Migration Review*, p. 105; EACH-FOR (2009). Environmental Change and Forced Migration Scenarios, Synthesis Report. http://rosamartinez.org/wp-content/uploads/2015/11/Migraciones-y-Cambio-Climatico_EACHFOR.pdf. Accessed 2 September 2018; Nansen Conference on Climate Change and Displacement in the Twenty-First Century (2011). Chairperson's Summary (para. 4). https://www.unhcr.org/4ea969729.pdf. Accessed 12 January 2020. With regard to legal studies only, see *ex multis*: Ntekangi (2014). *Vers un droit international des réfugiés écologiques*. Kinshasa: L'Harmattan; McAdam (2012). *Climate Change, Forced Migration, and International Law*. Oxford: Oxford University Press; Kälin & Schrepfer

G. Sciaccaluga, *International Law and the Protection of "Climate Refugees"*,
https://doi.org/10.1007/978-3-030-52402-9_2

how to interpret the phenomenon can indeed be quite relevant, there is a strong agreement on one crucial and basic element: climate change will play an ever-important role in determining and increasing migration flows worldwide, both at the domestic and international level.[3] Therefore, "even the most optimistic projections envisage a crisis of migration that will dwarf any we have seen so far".[4] Put simply, global warming increasingly affects the living conditions of different geographical areas of the planet, hence influencing both the quantity and the quality of migration patterns. This leads to an increase in voluntary and forced

(2012). Protecting People Crossing Borders in The Context of Climate Change: Normative Gaps and Possible Approaches. UNHCR Legal and Protection Research Series, PPLA/2012/01. https://www.unhcr.org/4f33f1729.pdf. Accessed 12 December 2016; Westra (2009). *Environmental Justice and the Rights of Ecological Refugees*. London: Earthscan; Xing-Yin Ni (2015). A Nation Going Under: Legal Protection for "Climate Change Refugees". *Boston College International & Comparative Law Review*, 2, pp. 329 ff.

[3] "It is expected that climate change in the 21st century will lead to an increase in the movement of people (medium evidence, high agreement)". IPCC (2014). *Summary for policymakers*. In Climate Change 2014: Impacts, Adaptation, and Vulnerability. Part A: Global and Sectoral Aspects. Work of Working Group II of the Fifth Intergovernmental Panel on Climate Change Report. Cambridge: Cambridge University Press, p. 20. See on the same topic but from another perspective also Bergoglio (2015). *Lettera enciclica – Laudato si' – del Santo Padre Francesco sulla cura della casa comune*. Vatican City: Libreria Editrice Vaticana, p. 23: "Climate change is a global problem with grave implications: environmental, social, economic, political and for the distribution of goods. It represents one of the principal challenges facing humanity in our day. Its worst impact will probably be felt by developing countries in coming decades. Many of the poor live in areas particularly affected by phenomena related to warming, and their means of subsistence are largely dependent on natural reserves and ecosystemic services such as agriculture, fishing and forestry. They have no other financial activities or resources which can enable them to adapt to climate change or to face natural disasters, and their access to social services and protection is very limited. For example, changes in climate, to which animals and plants cannot adapt, lead them to migrate; this in turn affects the livelihood of the poor, who are then forced to leave their homes, with great uncertainty for their future and that of their children. There has been a tragic rise in the number of migrants seeking to flee from the growing poverty caused by environmental degradation. They are not recognized by international conventions as refugees; they bear the loss of the lives they have left behind, without enjoying any legal protection whatsoever. Sadly, there is widespread indifference to such suffering, which is even now taking place throughout our world. Our lack of response to these tragedies involving our brothers and sisters points to the loss of that sense of responsibility for our fellow men and women upon which all civil society is founded".

[4] Behrman & Kent (Eds.) (2018). *"Climate Refugees" Beyond the Legal Impasse?* New York: Routledge.

movements of individuals and populations worldwide. In the present historical phase, it is almost concretely tangible that migration issues represent a very sensitive role in our personal and collective consciousness, as witnessed by the rise of nationalist and anti-migration political forces in many countries worldwide. And in a planet of exponential demographic growth stuck between ineluctable globalization processes and the need to preserve national/regional identities, the topic of "climate refugees" deserves special attention, since it affects a growing number of individuals, communities, and nations, putting under severe pressure the social, economic, and geopolitical balances of our entire international system.

As a premise, a brief look at the negative effects of climate change on ecosystems and ecosystemic services is helpful to better understand what we are dealing with. Fundamentally, "climate refugees" become such because of climate change-driven ecosystemic alterations that occur in their homelands. Desertification, sea-level rise, and the increase of dangerous meteorological events are climate-related processes that are deeply intertwined with ecosystems and their capacity to provide indispensable "goods and services" to humans. In the nineteenth century, the founding father of ecology, the German biologist Ernst Häckel, defined ecology as "the science studying the relationships between living organisms and the outside world, in which we can appreciate – generally speaking – the factors playing a role in the struggle for existence".[5]

It took several decades however for ecological studies and values to become internationally relevant not only from a scientific viewpoint, but also from a political and legal one. Ecological values and reasoning in fact started gaining international public attention in the last six or seven decades, namely, as a result of the widespread development of human societies, economic activities, and international trade that started to progressively pose existential threats to the environment and the planet as a whole. The advent of the industrialization era and the establishment of a global market has brought—alongside wealth, growth, and development—also sudden/local environmental man-made disasters, as well as slow-onset ones, such as desertification or rising sea levels, for instance. These "new" threats have led scholars, institutions and public opinion worldwide to attribute growing importance to environmental issues and values, and therefore to environmental and ecological studies

[5] In Déleage (1994). *Storia dell'ecologia. Una scienza dell'uomo e della natura*. Naples: Cuen, p. 61.

too. Notions and discoveries stemming from these fields have consequently experienced a growing dissemination, with social sciences starting to incorporate them within their own domains by virtue of one simple but important assumption: human and ecological systems have much in common.

A revolutionary contribution in this sense was offered during the Seventies by the English scientist James Lovelock, among the first to argue that the whole planet—Gaia, as he called it—in its organization, functioning, and structure has properties that are very similar to a living organism.[6] His ideas had a first impact on the intellectual level, fostering studies, debates, and discussions that had in the end a significant scientific outcome as well. Lovelock put forward the hypothesis that as individual organisms have the ability to self-organize and control certain parameters (for example body temperature and blood chemistry), so the global ecosystem network has the ability to regulate, at least partially, the average global temperature, the chemical composition of the atmosphere, and so on. The extreme novelty of his thesis led initially to trivial and simplistic interpretations. However, thanks to subsequent research and the introduction, for instance, of concepts such as the dissipative structure,[7] the "Gaia theory" has gained strong scientific support. If accepted, it entails radical outcomes both ethically and politically, since it demonstrates the existence of a climatically and environmentally interdependent world, on which the very ability for human societies to develop and grow depends in the long run.

Not surprisingly, the very basic elements of the Gaia theory can be found today in the good part of the political and legal instruments adopted since the birth of international environmental law. The 1972

[6] Lovelock (1979). *Gaia: A New Look at Life on Earth*. Oxford: Oxford University Press.

[7] A dissipative structure (or dissipative system) is a thermodynamically open system that works in a state far from thermodynamic equilibrium, exchanging energy, matter, and/or entropy with the environment. Examples of dissipative structures include phenomena such as cyclones and, on a larger and more complex scale, ecosystems and life forms.

Stockholm Declaration on Human Environment[8] or the 1994 Draft Principles on Human Rights and the Environment,[9] for instance, clearly reflect the idea that the international community is environmentally interdependent, and that it is a community not to be exclusively considered in an anthropocentric way. More recently, the same assumptions have constituted the basis of the 2030 Sustainable Development Goals,[10] which outline the objectives that UN Heads of State and Government have set for 2030. Similar evolutions can be retrieved in the field of international law too. According to some scholars, for example, given the trends of the current globalization era, we should talk today of "international interdependence law": if after the 1648 Peace of Westphalia and the birth of the modern international state system, international law aimed essentially at granting basic rules for the coexistence of such entities (constituting therefore an "international coexistence law"), nowadays, after almost four centuries, the international legal system has evolved up to the point of governing not the mere coexistence between states, but rather the interdependent system in which they act.[11]

At a deeper level, this reasoning also reflects some ancient philosophies that explain the common origin and interdependent functioning of all things. The seminal Taoist book *Tao Te Ching* solemnly recites, for instance: "The Way gave birth to one. One gave birth to two. Two gave birth to three. Three gave birth to all things". More importantly however from a scientific viewpoint, this perspective also reflects the development experienced by human knowledge in the last century, a development that is still ongoing not only because of ecological studies, but, more significantly, because of the major discoveries that modern physics brought to light during the twentieth century.[12] It is common knowledge for

[8] UN Declaration on Human Environment (Stockholm Declaration), adopted 15 June 1972.

[9] Draft Principles on Human Rights and the Environment, adopted 16 May 1994, E/CN.4/Sub.2/1994/9, Annex I.

[10] UN General Assembly Resolution (2015). Sustainable Development Goals: Transforming Our World: The 2030 Agenda for Sustainable Development, UN Doc. A/RES/70/1.

[11] See in this regard: Picone (2013). *Comunità Internazionale ed Obblighi Erga Omnes*. Naples: Jovene Editore, p. 16. See also: Giuliano, Scovazzi & Treves (1983). *Diritto internazionale. Vol. 2: Gli aspetti giuridici della coesistenza degli Stati*. Milan: Giuffré.

[12] Capra (1975). *The Tao of Physics*. Boston: Shambhala. See also Kaiser (2011). *How the Hippies Saved Physics*. London: W. W. Norton.

example that Einstein's relativity theory, Heisenberg's indetermination principle, or Bell's theorem on quantum entanglement have shown how it is completely misleading (and in fact simply incorrect) to believe in the existence of closed, separated, and independent "units": everything, even space and time, and therefore quite obviously also nations, states, and human communities, are interrelated with the system in which they live and with the other actors they interact with.

With the years, the notion of interdependence, on which the entirety of ecology is based, has started to be incorporated within social sciences and, among them, within international law, too. As a matter of fact, it is probable that only through the acceptance of interdependence as a basic international law concept and principle one may hope that the "common heritage of mankind", namely, oceans, biodiversity, forests, and the stability of the climatic system we live in, may be effectively preserved in the future. To this end, an international law based on interdependence seems more and more needed in the light of globalization and of the fourth industrial revolution, that is putting in crisis the "classic" international (law) system, which is deeply anchored to the notion of sovereignty. In a globalized and interconnected world, where the movement of people, information, financial and natural resources has dramatically increased and accelerated, states and, more generally, international legal subjects cannot be considered (and should not consider themselves) as independent, absolute, and stand-alone units.

In this reasoning, the concept of ecosystem is of primary importance. An ecosystem consists of all the living organisms and non-living factors that are present in a certain environment, and—equally importantly—of the relationships that link these different elements. A more detailed definition is adopted by the 1992 Convention on Biological Diversity,[13] according to which an ecosystem is "*a dynamic complex of plant, animal and micro-organism communities and their non-living environment interacting as a functional unit*". When we consider the environment therefore, we must take into account not only its value as an order or a constraint, but also as an organization.[14] In this sense, the environmental degradations we are living through today should not only

[13] Convention on Biological Diversity, adopted on 22 May 1992, entered into force 29 December 1993.

[14] Bagliani & Dansero (2011). *Politiche per l'ambiente. Dalla natura al territorio.* Novara: Utet, p. 32.

be considered as a mere loss of natural resources, but—more interestingly—also as a change of the systems and the dynamics we live in. In fact, at the very basic level, our ability to develop as human communities or societies depends on the "services", "goods", and "functions" that ecosystems continuously offer us. This is of course no novelty.

The appreciation of human dependence on nature has always been present throughout humankind's history. Indeed, it was probably much clearer in the past when this bond was more direct and visible to the majority of people. Today, with the industrialization of food production systems, more and more individuals, especially in developed states, rarely get in touch with the activities that provide them with food, such as farming or animal slaughtering, for instance. Back in the past, these activities were evidently more visible to people generally, something that made them more aware of their dependence on natural resources. No surprise then that Plato expressed his deep concern about the excessive deforestation of the Attic region in ancient Greece, which was leading to excessive soil erosion and to the drying up of water supplies.[15] Similarly, ancient civilizations in China, India and South America were well aware that the excessive exploitation of land endangered food production.[16] In the last decade, a great scientific contribution to better understanding this field was offered by Jared Diamond. In his enlightening work "*Collapse: How Societies Choose to Fail or Succeed*", Diamond shed light on those cases in human history in which certain societies have collapsed or dramatically receded because of (gravely) unwise and short-sighted exploitation of the natural resources they depended on.[17] Spanning from the Vikings in Greenland to Polynesian societies in the Pacific, Diamond's book shows that a significant misuse of the resources a community depends on may condemn it to decadence, weakness or even to collapse. His studies probably deserve attention in any work dealing with anthropogenic global warming, considered its capability to alter the balance and the functioning of our entire planet. In fact, due to its inertial nature, worldwide diffusion, and expected worsening, global warming may radically change the

[15] In Daily (Ed.) (1997). *Nature's Services: Societal Dependence on Natural Systems*. Washington, DC: Island Press, p. 5.

[16] Donald Hughes (1975). *Ecology in Ancient Civilizations*. Albuquerque: University of New Mexico Press.

[17] Diamond (2005). *Collasso. Come le società scelgono di morire o vivere*. New York: Viking Press.

environmental landscape in which mankind has evolved[18] in relatively stable climate conditions since the start of Holocene, around 10,000 years ago.[19] Climate change has—sobering to recall—an inertial and long run intrinsic nature: the average atmospheric lifetime of a CO_2 particle, the most widely diffused greenhouse gas (GHG), spans from 50 to 200 years on average,[20] meaning that even in the absurd hypothesis of a radical and immediate reduction of greenhouse gas emissions worldwide, the negative effects of global warming will keep worsening in the next decades. For instance, the recent 2020 COVID-19 pandemic shows that some *force majeure* events can potentially radically decrease GHG emissions worldwide. Nevertheless, even in the presence of phenomena of this magnitude and effect, global temperatures will keep increasing. This is why future measures of "climate engineering", such as greenhouse gas capture from the atmosphere and solar radiation management, appear today inevitable.[21]

In this context, if we trust climate science, only through a careful and decisive management of climate change, both from a mitigation and adaptation perspective, we may hope that our societies will have the

[18] It is sobering to recall that in the twentieth century alone the human population has quadrupled, while the world economy and industrial production have increased, respectively, by 14 and 40 times. See in this regard, *ex multis*: McNeill (2010). *Du Nouveau sous le Soleil: histoire de l'environnement mondial au XXe siècle*. Seyssel: Champ Vallon, p. 44.

[19] The Holocene, the last and current geological era, started between 12,000 and 10,000 years ago after the end of the last ice age. The Holocene has allowed, also for its unusual climatic stability, an enormous human development, with the settlement of sedentary human societies and agriculture. For more information, see: Incropera (2016). *Climate Change: A Wicked Problem*, Chapter 2. Cambridge: Cambridge University Press

[20] IPCC (2016). Fifth Report, Working Group I, The Scientific Base. IPCC. https://archive.ipcc.ch/ipccreports/tar/wg1/016.htm. Accessed 13 January 2020.

[21] On which see: Parson (2017). Opinion: Climate Policymakers and Assessments Must Get Serious About Climate Engineering. *Proceedings of the National Academy of Sciences*, 35. https://www.pnas.org/content/pnas/114/35/9227.full.pdf. Accessed 4 August 2019. See also: Parson (2017). Starting the Dialogue on Climate Engineering Governance: A World Commission. Fixing Climate Governance Policy Brief No. 8. Center for International Governance Innovation. https://www.cigionline.org/publications/starting-dialogue-climate-engineering-governance-world-commission. Accessed 4 August 2019.

opportunity to prosper in future.[22] This only appears possible in case of success in the preservation of those climatic and environmental balances that have been more or less stable since the start of Holocene, which allowed the first sedentary human settlements and the agricultural revolution. Today, these balances are existentially threatened in many areas of the world. The warming of average global temperatures, followed by sea-level rise, acidification of oceans, modifications of carbon release and capture processes, alongside with desertification, can largely modify the environmental and climate system on which the well-being and the good functioning of human societies worldwide depend.

This is not the place to focus in depth on the hard science concerning climate change. However, some information is necessary to correctly understand the magnitude of the phenomenon, and hence to better conceive the whole issue of "climate refugees". First of all, from the point of view of the novelty and significance of what humanity has begun to face, it is important to underline that concentrations of greenhouse gases in the atmosphere (among which CO_2, CH_4, and N_2O play a primary role) have never been so high in the last 800,000 years (and perhaps even in the last two million years). Focusing only on the last 20,000 years, however, the average concentration of CO_2 equivalent particles in the atmosphere has been—with some slight fluctuation—around 270 parts per million (ppm). Today, despite all efforts and good intentions, we have passed 400 ppm, and this number is steadily growing.

As briefly outlined above, due to their inertial trend, the climate modifications that are in place today will continue impacting on our ecosystems for decades, threatening those environmental balances that offer (not only) to mankind all the "goods" and "services" on which we have developed as societies in the last millennia. Ecological studies provide us with quite an exhaustive categorization of the goods and services that ecosystems offer.[23] Generally, ecosystem services refer to the conditions and

[22] See Nuwer (2017, April 1). How Western Civilisation Could Collapse. *BBC online*. https://www.bbc.com/future/article/20170418-how-western-civilisation-could-collapse. Accessed 13 June 2029.

[23] Millennium Ecosystem Assessment (2005). *Ecosystems and Human Well-being: Synthesis*. Washington, DC: Island Press. See also: Secretariat of the Convention on Biological Diversity (2016). Managing Ecosystems in the Context of Climate Change Mitigation: A Review of Current Knowledge and Recommendations to Support Ecosystem-Based Mitigation Actions That Look Beyond Terrestrial forests. CBD Technical Series No. 86. https://www.cbd.int/doc/publications/cbd-ts-86-en.pdf. Accessed 12 May 2019.

processes through which ecosystems and their components sustain human life. These services "produce" different sorts of vital goods, such as food, natural fibers, biofuels, essences, pharmaceutical principles, and others. Moreover, ecosystems do not only offer "goods", but they also provide all those functions that are necessary for the creation and the development of life: for instance, the production of oxygen, the purification of water, pollination, the partial control of biogeochemical cycles,[24] and the recycling of nutrients and biodegradable substances.

Hence, the link between climate-driven environmental degradation and human well-being becomes vividly observable from both an individual and a collective perspective. The current ecological crisis, which is due to climate change but more generally to the excessive exploitation and pollution of land, air, water, and natural resources, negatively impacts on the ability of ecosystems to offer us indispensable natural goods and services in the medium and long run. Rising quantities of greenhouse gas in the atmosphere affect carbon dioxide sequestration processes, and, thus, climate patterns worldwide, with the consequent rise of the global average temperature. Global warming, in turn, through phenomena like desertification and sea-level rise leads in some world regions—that are often densely populated and scarcely prepared from an institutional, social, and economic point of view—to reductions in food, natural fibers, and fuels. These changes can furthermore lead to the potential loss of important cultural values, characterizing, in their infinite diversity, the global human context. Some micro-state islands in the Pacific and Indian Oceans, for instance, could see their territory completely submerged by rising seas in the next decades, leading not only to inevitable "material" losses, but also to the loss of all the symbolic and emotional elements that bind a population to its territory.

The most invasive impacts caused by climate change deal and will deal with water availability[25]: on the one hand, some regions experience a reduction in water supply (desertification, heat waves, droughts), especially at the Tropics, in the Mediterranean, in the Middle East, and at the

[24] The term biogeochemical cycle indicates the circulation of chemical elements between living and non-living components. Since this cycle involves both ecosystems and geological processes (water movement, erosion, etc.) and chemicals, the resulting cycles are called biogeochemical.

[25] IPCC (2007). Climate Change 2007: Synthesis Report, pp. 30 and 53. https://www.ipcc.ch/site/assets/uploads/2018/02/ar4_syr_full_report.pdf. Accessed 12 May 2019.

southern limits of Africa and Latin America; on the other hand, excessive water availability (meaning heavier rains and floods) affects parts of Eastern Africa, India, China, and northern latitudes in general. Moreover, due to rising sea levels, the densely populated areas of the big river deltas in the world (such as the Ganges in India, the Mekong in Vietnam, the Nile in Egypt and the Yangtze in China), not to mention the atolls standing out only a few meters above sea level, will increasingly be subject to floods, rising tides, salinization of water resources and soils, degrading coral reefs, and territorial submersion.

Unsurprisingly, these environmental changes have already started affecting migration flows in different parts of the globe. Where environmental degradation affects areas that are institutionally and socioeconomically fragile (as it is often the case in developing countries), they exacerbate migratory push-factors, inducing more people to leave their homes and migrate somewhere else, be it within their state or abroad. It is therefore quite clear that global warming, moreover when associated to worldwide exponential demographic growth, leads to more migration because it worsens living conditions. If this is true, it is however not clear from an international law perspective who, among those who migrate, should enjoy international protection and who should not. In fact, the well-known differentiation between "economic migrants" and "refugees" plays a euphemistically crucial role also with regard to climate-driven migrations, notwithstanding their novelty and peculiarity.

In the following chapters, attention is dedicated to the different and sometimes contrasting attempts to conceptualize the category of "climate refugees", with the general aim of identifying and defining which of them could or should be granted the right to international protection when migrating abroad.

PART I

Are All Climate Change Migrants Equal? Making Light in the "Climate Refugees Cauldron"

Introduction

When compared to the history of human migration, the phenomenon of "climate refugees" is a newborn. Because of its novelty and intrinsic complexity, there are several different ways to conceptualize or interpret the phenomenon as a whole, not only at the political and media level, but also within academia. Depending on the approach one follows, significantly different outcomes emerge with reference to the number of people involved and, consequently, also to the political and legal solutions proposed to cope with this challenge. This often leads to a confused "climate" that surrounds the very nature of the topic. After all, however, "in a new field unsure of its identity a certain amount of conceptual and semantic chaos is unavoidable".[1]

Therefore, the debate on climate-induced or forced migrations—on "climate refugees"—is still open and sometimes even incoherent after almost four decades of dedicated studies. Radically different narratives through which one may look at the topic still emerge in the public discourse. For example, the phenomenon of "climate refugees" can be looked at through a "scientific", a "capitalist", a "humanitarian", or a

[1] Merryman (1977). Comparative Law and Social Change: on the Origins, Style, Decline & Revival of the Law and Development Movement. *American Journal of Comparative Law*, 25, p. 457 ff, cit. in Mayer (2012). "Environmental Refugees"? A Critical Perspective on the Normative Discourse. Center for Sustainable Development Law. https://papers.ssrn.com/sol3/papers.cfm?abstract_id=2111825. Accessed 14 January 2016.

"radical" discourse, each of these entailing different assumptions, interpretations, and research outcomes.[2] Thus, "the debate on the topic sees the convergence of conflicting discourses [...] onto apocalyptic narratives that forecast massive, abrupt and unavoidable flows of climate refugees".[3]

Further upstream, these diverging conceptualizations largely depend, basically, on the scientific predictions on which one relies in relation to global warming. It is crystal clear that the density and intensity of climate-related migrations varies depending on the increase in the average global temperature that will effectively occur in the next decades due to the greenhouse effect. To date, the international community identifies in a 2°C increase compared to 1850 the threshold below which the negative effects of climate change can be contained within acceptable and sustainable boundaries. With a 2°C increase by 2100 there should be the possibility to keep the socio-political and socio-economic order sufficiently stable and secure at a global level. Beyond this point, in Diamond's words, we risk the collapse of the international system as we currently known, overwhelmed by a rising tide of unmanageable human migrations and conflicts. In reality, the Paris Climate Agreement, which entered into force into 2016 and has to date been ratified by 189 parties, sets the preferable objective of containing global warming to 1.5°C, and this mainly to preserve the territorial integrity of low-lying insular states that risk being submerged by rising sea levels. However and unfortunately, according to current estimates (and remembering that the increase in global temperatures that have already occurred since 1850 coincides roughly with 1°C), it is estimated that without decisive and radical mitigation interventions, a rise of well over 3°C will be achieved by 2100. And it is also useful (and worrying) to recall that the IPCC and climatology also consider the possibility of other scenarios—that are unlikely but not impossible—of an increase up to and above 6°C by the end of this century, with the consequent radical collapse of the climatic and ecosystem balances we all rely on, and perhaps the extinction of humankind.

[2] For further information, see: Bettini (2013). Climate Barbarians at the Gate? A Critique of Apocalyptic Narratives on "Climate Refugees". *Geoforum*, 45, p. 63 ff. https://www.sciencedirect.com/science/article/abs/pii/S0016718512001947?via%3Dihub. Accessed 12 January 2020.

[3] *Ibidem*.

Consequently, when we think of "climate refugees", it is then clear that the humanitarian, political, and economic relevance of the phenomenon significantly changes depending on how and how much the global average temperature will effectively rise in future. Without serious climate change mitigation policies, it will be extremely difficult to manage and control climate-related human movements, with ever-greater internal and international migrations induced by the excessive deterioration of socio-economic conditions in many parts of the globe, especially in the poorest and most vulnerable ones. Conversely, were the international community to act decisively against climate change from a mitigation viewpoint, international migration policies, and international protection systems could be enacted within the boundaries of social and political sustainability.

While in practice there is absolutely no certainty that the Paris Agreement's goals will be met, from a normative standpoint, the main argument of this book hinges on the hope and assumption of a 2°C increase. This is because any proposed management or solution to the issue of climate-related migrations only makes sense if the Paris Agreement's goals are met. If states fail in containing global warming within ecosystemic (and hence socially) sustainable boundaries, there is virtually no chance that the international legal order will be able to cope with the phenomenon here at stake in a dignified, decent, and human way.

To date, the fact is that even if the most radical greenhouse gas mitigation policies are enacted, global warming has started, that it will last long and hence that, even in the best conceivable scenario, migration flows will increase, triggering different reactions by states and the international community to ensure their management. In a nutshell, therefore, depending on the scientific predictions followed and on the degree of trust in the international community's willingness and ability to tackle climate change, the whole "climate refugees" picture can radically change.

Actually, it is possible to reduce the different existing conceptualizations to two broad categories: maximalism and minimalism. The first one is offered by maximalists,[4] who were the first ones to study climate-driven migrations, between the end of the Eighties and the beginning of the 2000s. Maximalists were mainly environmental sciences scholars, the first academic group that had the chance to appreciate the significant link existing between anthropogenic climate change and human

[4] Cfr. in this sense Jacobson (1988). Environmental Refugees. A Yardstick of Habitability. *Bulletin of Science, Technology and Society* 8, p. 257 ff.

migrations. Quite reasonably, due to their knowledge and background they very often gave too little importance to migration and international law studies, tending eventually to alarmist and public impact tones through exorbitant numerical forecasts concerning the numbers of people potentially involved. In reaction to maximalism, the minimalist school started, from the beginning of the Nineties, to respond to the catastrophist figures and theses suggested by excessively alarmist scholars.[5] In the years that followed, minimalism consolidated a scientifically more mature line of research that has progressively become widely accepted within the academic community. In comparison to maximalism, minimalism follows a more sober approach, based on empirical studies on migration and more anchored to the existing regulatory frameworks and, therefore, in particular to existing international protection instruments. Over the years, minimalism has thus succeeded in avoiding some of the excessive simplifications that are typical of (part of) the maximalist school.

Studies investigating the causal relationships between climate and/or environmental factors and related migrations have amply demonstrated how this relationship is extremely complex, and that it does not follow a linear trend. This means that climate change cannot be easily considered as the sole or principal cause of the displacement/migration of individuals. It follows that it is too simplistic to think of masses of people forced to move only due to the adverse effects of climate change in their countries of origin. However, such an image is certainly appealing from a media viewpoint: by highlighting the link between migration and climate, and thus implicitly between climate and security, it helps to denounce and highlight the negative effects of global warming and therefore to trigger a reaction. This securitization of climate through its nexus with migration can thus be viewed as having political and policy upsides when viewed from the normative angle of combating climate change. Nonetheless, excessive catastrophism can have negative consequences too: this especially in the political domain, where the predictions of "hordes" of

[5] Bilsborrow (1992). Rural Poverty, Migration, and the Environment in Developing Countries: three case studies, World Bank Policy Research Working paper 1017. World Bank. http://documents.worldbank.org/curated/en/776914687673865l6/pdf/multi0page.pdf. Accessed 14 May 2019. See also: Black (1998). *Refugees, Environment and Development*. New York: Harlow Longman; Kibreab (1997). Environmental Causes and Impact of Refugee Movements: a Critique of the Current Debate. *Disasters*, 21, p. 20 ff.

desperate climate migrants, can—in addition to solidarity and humanitarian feelings—provoke closure and rejection at the national or regional level, and thus exacerbate existing tensions.

In this sense, Ulrich Beck's argument resonates. The German sociologist argues that global warming already has a decisive impact on human societies not so much because of its physical characteristics (temperature increase, sea-level rise, variation of ecosystem balances, etc.), but rather because of its symbolic components and, therefore, because of the repercussions that impact on social and political balances worldwide already today.[6]

On the sidelines of these considerations, there is a further negative effect driven by an excessive dramatization of the phenomenon of climate refugees. In fact, improperly defining "climate refugee" any migrant, whether forced or voluntary, internal or international, that is subject to climate pressures is, from a legal perspective, not only incorrect, but it also creates confusion and may undermine existing protection systems.[7] In public discourse, the question of "climate refugee" still reflects the conclusions of early maximalist studies because of their media-prone tone. In this perspective, "climate refugees" are usually seen as individuals:

i. forced to abandon their homes due to climate change,
ii. they will be hundreds of millions,
iii. and they will cross international borders *en masse*.

More careful research shows, instead, that

i. the causal link between climate change and migrations is important but not all-encompassing,
ii. most of the predictions relating to the people involved are in the best hypothesis a mere "guess",[8]

[6] Beck (2011). *Disuguaglianza senza confine*. Bari: Editori Laterza.

[7] Such a risk is conceivable because, if the international law instruments in charge of refugee protection were modified to also include "climatic refugees", many states could refuse, even for long periods, to adhere to the modifications or to the instrument in full. In this case, the entire system for the protection of refugees, whose weaker side is its poor implementation within state parties, could be seriously undermined.

[8] "The scope and scale of displacement due to slow-onset disasters, such as drought, remains a guess at best", in IASC Task Force on Climate Change (2010). Quick guide to Climate Change Adaptation. IASC. https://interagencystandingcommittee.org/system/files/legacy_files/IASC%20Quick%20Guide%20Adaptation.pdf. Accessed

(iii) most people involved migrate within their national boundaries, thus not representing an immediate international (law) challenge.

Rough beliefs about the topic often derive from an oversimplification of reality. In this sense, they are of limited use in informing consequent policy arguments. A careful and scientific analysis, based on the study of existing migration flows and on empirical research carried out in this field, even if less clamorous and more gradual, has the merit of directing policy-makers toward solid paths. Future systems of governance and protection of the rights of "climate refugees" ought to rely on such work.

A consequent objective of the first part of this book is to identify which "climate refugees" should be protected under international law. The existence of different approaches and the proliferation of many and different definitions of the category has, in fact, led to the creation of a "climate refugees cauldron" within which any person whose migration is directly or indirectly linked to climate change can be effortlessly thrown. The law has to trace some lines somewhere to enjoy a certain degree of effectiveness. In relation to "climate refugees", it has to find and set criteria through which it is possible to identify which individuals are actually susceptible of being protected through international law and which not: in this sense, some light must be shed onto this "cauldron", with "climate refugees" that unavoidably emerge as a conceptually and semantically very confused category. The problem of the distinction between "economic migrants" and "refugees" (namely, between those who do not enjoy an effective right to assistance and protection in a third state and those who do) is always of utmost importance in any international protection reasoning. Climate or environmental refugees are no exception: in the name of worsened environmental conditions in different areas of the world, one could, pushed by moral and ethical principles, perhaps believe that all those who migrate or will migrate because of global warming should intrinsically be endowed with a right to international protection. In an international law perspective, this is not and cannot be the case.

12 May 2019. See also: OCHA, IDMC & Norwegian Refugee Council (2009). Monitoring Disaster Displacement in the Context of Climate Change: Findings of a Study by the United Nations Office for the Coordination of Humanitarian Affairs and the International Displacement Monitoring Center. https://www.internal-displacement.org/sites/default/files/publications/documents/200909-monitoring-disaster-displacement-thematic-en.pdf. Accessed 14 November 2019.

The following pages provide the reader with a conceptualization of "climate refugees" useful not only for the purpose of a proper understanding of the phenomenon but also for eminently legal purposes. We will try to understand if, within the "climate refugees cauldron", there are any particular individuals that could or should be protected under international law when they migrate and seek protection abroad.

CHAPTER 3

Climate Change and Migration

Climate change and human migrations are two delicate fields of study. It is difficult to make sense of the two and the relationships between them. After all, they both try to understand two of the most complex and mysterious fields of reality: the nature of climate and the behavior of human beings. Even though scientific progresses in climatology have been astounding in the last decades, especially after the political push generated by global warming, many of the conclusions reached remain uncertain. For example, when talking of sudden-onset disasters such as floods or windstorms, asking if climate change is the cause of a particular event is a meaningless question, that "cannot be answered[1]". Climate science tells us that the relationship between global warming and a single extreme meteorological event can only lead to a probabilistic and not a binary attribution.[2] By affecting the probability of certain weather patterns, excessive greenhouse gas "indirectly affect social systems in diffuse and amplifying ways with consequences which, like the concentric

[1] Huber & Gulledge (2011). Extreme Weather and Climate Change: Understanding the Link and Managing the Risk, p. 2. Center for Climate and Energy Solutions. https://www.c2es.org/site/assets/uploads/2011/12/white-paper-extreme-weather-climate-change-understanding-link-managing-risk.pdf. Accessed 13 October 2019.

[2] Pall et al. (2011). Anthropogenic Greenhouse Gas Contribution to Flood Risk in England and Wales in Autumn 2000. *Nature*, 470(7334), p. 382.

G. Sciaccaluga, *International Law and the Protection of "Climate Refugees"*,
https://doi.org/10.1007/978-3-030-52402-9_3

circles that an impact produces on a water surface, extend ad infinitum and *at absurdum* in time and space[3]". Interestingly in this sense, climate change is often defined as a "super-wicked" problem, something that due to its interdependency, circularity, multi-causality, and multi-effects remains uncertain in its very nature, and consequently extremely difficult to be dealt with.[4]

If this is true when we talk of climate-related processes or events, it is even more so when we look at human migrations. The choice to migrate is in fact always determined by a complex set of factors. An individual can be pushed to leave his or her country for different reasons: climatic, environmental, economic, demographic, familial, social, psychological, and so on. In turn, the same factor can have a different effect depending on the person it affects. It is often difficult to determine which factor has played the decisive role in leading a person to leave his or her home, and this is because different persons living in the same context can react differently to similar challenges. Just imagine, for example, the case of a young Filipino man, who, without deep family ties and with enough resources at his disposal, decides to emigrate abroad in search of fortune *also* because of the frequent cyclones and floods that hit his area of origin. At the same time, think of a Filipino mother linked to her maternal duties who, in the same socio-environmental context, cannot emigrate and decides to stay. Likewise, climate or environmental degradations that are similar to one another but that take place in different places can lead to different effects in relation to the individual choice to emigrate. Just as a matter of example, peasants in southern Italy facing heat waves and drought, when compared to farmers in Mali facing similar environmental conditions, do not suffer the same way, because their socio-economic and political-institutional *milieu* is radically different. Hence, comparable climate-related environmental degradations have different impacts

[3] Mayer (2017). Climate Change, Migration and the Law of State Responsibility. In Mayer & Crépeau (Eds.), *Research Handbook on Climate Change, Migration and the Law*. Cheltenham: Elgar, p. 256.

[4] Lazarus (2009). Super Wicked Problems and Climate Change: Restraining the Present to Liberate the Future. *Cornell Law Review*, 94, p. 1153 ff.

on similar professional categories depending on their resilience capabilities.[5] In a hypothesis such as this one, it would be then possible to argue that migrations linked to drought, heat waves, and global warming in Mali are primarily due to poverty and institutional failures, rather than to climate change alone. In a nutshell, the decision to migrate—even in the realm of forced migrations—is multifaceted and dispels common assumptions about linear cause and effect relationships.[6]

It is thus difficult to establish a precise etiological link between climate change and human migrations. However, part of the literature has ignored this important aspect, and anthropogenic climate change has often been viewed as the sole cause of displacement of entire populations. Theoretical and empirical research on migration show though that the impact deriving from climate factors is not only mediated by the degree of vulnerability, preparation, and resilience of the affected communities, but also by already established migration flows and schemes, and from the potential forms of adaptation existing *in situ* as well. So, although it is indisputable that climate factors have (and will increasingly have) a significant role in determining migratory flows, it is equally true that several other push-factors must be taken into account. A remarkable and commonly accepted interpretation is thus to consider climate change as a "threat multiplier": something that exacerbates already complicated situations. In this sense, in her milestone book *Climate Change, Forced Migration, and International Law* Jane McAdam reports the thoughts of a Bangladeshi public official, according to whom poor development, combined with the over-use of cultivable areas, demographic pressure, deforestation, and settlement of populations in environmentally fragile areas, means that people living in such situations are intrinsically vulnerable to natural disasters. Through this approach, climate change, through its negative impacts on the quality of the environment, can become that factor that drives certain individuals to emigrate.[7]

[5] The same reasoning is proposed in Meze-Hausken (2000). Migration, Development, and Environment: How Vulnerable Are People in Dryland Areas? A Case Study in Northern Ethiopia. *Mitigation and Adaptation Strategies for Global Change*, 5, p. 379.

[6] Hugo (1996). Environmental Concerns and International Migration. *International Migration Review*, 30, Chapter. 6.

[7] In similar terms, see also Mayer (2018): "The decision of individuals to migrate has a lot to do with the costs and benefits associated with non-migration decisions. This inherent link between migration and other issues of national development makes it

Overall, the choice to migrate depends on a long list of factors—on which climate change has an impact—but among which it is difficult to establish a clear hierarchy.[8] Among the various causes, we may for example consider the political and institutional context,[9] the environmental situation, the size of the family unit, the economic situation in the place of origin and the potential destinations, the demographic pressure *in situ*, education opportunities, the presence or otherwise of common social networks in potential destinations, and so on. This means that we are dealing with a quite complex context to be clearly understood. Then, a further element that is rarely considered is that migrations can frequently bring positive effects on the communities of origin of migrants. Several studies show how remittances constitute an extremely positive element for the countries of origin of migrants, that are often very poor and in urgent need of external resources.[10] Furthermore, the emigration of a family member can be experienced as a form of positive adaptation to a difficult socio-economic-environmental context. Circular and seasonal migrations, typical of some African regions for example, as well as permanent ones, lighten the economic weight of a family unit and, at the same time, often allow an additional revenue for the family itself thanks to the sending of remittances.

Similar considerations of course show that migratory processes should, like anything else, be read both in their positive and negative lights. Excessively simplistic readings of the phenomenon that foment alarmist figures risk leading to undesired reactions also from the point of view of human rights protection. In other words, to imagine climatic migrants as hordes

relatively perilous to seek on-size-fits-all 'solutions' to 'climate migration' in isolation from careful consideration for development in particular contexts". Mayer, *Climate Change, Migration and the Law of State* responsibility, in Mayer & Crépeau (Eds.). *Research Handbook on Climate Change, Migration and the Law*. Cheltenham: Elgar, p. 241.

[8] Hugo (2010). Climate Change-Induced Mobility and Existing Migration Regime in Asia and the Pacific. In McAdam (Ed.), *Climate Change and Displacement, Multidisciplinary Perspectives*, Oxford: Hart Publishing, p. 9 ff.

[9] The importance of the socioeconomic and political context is also underlined by Amartya Sen, who underlines that never a serious famine has hit a country with a democratic government and free press. See Sen (1993). Overcoming Global Hunger, Actions to Reduce hunger worldwide. Report. http://www.nzdl.org/gsdlmod?e=d-00000-00---off-0hdl--00-0----0-10-0---0---0direct-10---4-------0-1l--11-en-50---20-about---00-0-1-00-0--4----0-0-11-10-0utfZz-8-00&a=d&cl=CL2.22.2&d=HASH0144a39d21fe11a28a32cda9.12.fc. Accessed 10 March 2020.

[10] See as a matter of example: Barnett & Webber (2010). Migration as Adaptation: opportunities and Limits. In McAdam (Ed.) (2010) (op. cit.), p. 37 ff.

or tides of poor people in search of help in foreign countries can induce some national governments to respond to the intensification of migratory flows by seeking borders' closures, with the establishment of immigration limits and, therefore, through the rejection of the principle/duty of international cooperation existing in the field of humanitarian protection. This is essentially why it is necessary to adopt an approach that is as objective as possible, so as not to feed sensationalism based on superficial conceptualizations.

Further on, if it is very difficult to determine a direct causal link between climate change and migration, it is equally complicated to clearly differentiate between voluntary and forced migration. The threshold beyond which migration becomes forced and, therefore, below which it can still be considered voluntary is often blurred. It is indeed increasingly difficult to make this distinction. If, as a matter of example, a person escapes persecution in Eritrea and ends up in Tunisia, where he looks for a job but cannot find one, and after some years he arrives to Spain, what is this person to the Spanish state? There is of course a boiling political and ethical debate on these issues, but from a mere legal (and practical) viewpoint, it is crucial to maintain anyway the distinction between voluntary and forced movements. If the distinction were eliminated, there would be the possibility to impede the access to international protection standards to everyone, because, in the very end, the idea that "everyone is no one" partially reflects the truth. Therefore, from the point of view of protecting the persons involved (and also for the manageability of reception and assistance policies in the host countries), this distinction is crucially important.

Voluntary/preventive migrants, although pushed by climate factors, can only be included in the category of those who decide to leave their country mainly for economic opportunity reasons. These persons will not enjoy particular international protection regimes abroad, since their migration is not strictly speaking forced. On the other side of the *spectrum*, those who are actually forced to migrate internationally in the face of a climate change-related disaster that makes their country of origin uninhabitable should presumably benefit from higher international protection standards. They could hence, in principle, be considered as "real climate refugees": individuals who would face serious violations of their basic human rights in their origin country without the assistance of a third state or the international community as a whole.

It is therefore useful to think of climate migrations—both voluntary and forced ones—along a *continuum*. A clear line of separation between the two extremes cannot be easily determined, above all with regard to those climate or environmental disasters that occur gradually over time, such as, for example, sea-level rise. In fact, while it may be easier to determine the degree of "forcedness" of a migration cause by a sudden and violent event, it is more difficult to understand when and how a migration actually becomes forced (especially in a short-term perspective) in presence of slow environmental degradations. A citizen of the insular micro-state of Kiribati may more or less voluntarily choose, even today, to leave his or her country also due to the expectations concerning sea-level rise and the possible submersion of his/her territory. At the moment, this migration is then a voluntary one, not likely to activate any form of particular protection stemming from international law provisions, but after what specific date and corresponding sea-level rise should such migration be considered forced? In fact, cases of this kind have already happened, with some Australian and New Zealand tribunals expressing enlightening viewpoints. By way of example, we can think of a New Zealand court,[11] which, judging on an asylum plea of an I-Kiribati, refused to recognize the refugee status to the applicant on the basis of the 1951 Geneva Convention Relating to the Status of Refugees[12] (hereinafter referred to as the 1951 Refugee Convention). According to the judges, the applicant was not suffering in the country of origin, at the time of the asylum plea, of environmental conditions capable of integrating a persecution as understood under international law.

Nevertheless, the problem persists: if predictions regarding the rise of the oceans for the coming decades should prove to be exact, a state like Kiribati (and with it several others) could even see their territories completely submerged by the end of the century. The populations of these countries will be forced to gradually leave their homes and to seek some form of protection and assistance in other states, therefore also relying on international law. Hopefully, at that point the international community

[11] New Zealand Immigration and Protection Tribunal (NZIPT) Decision 800413/2013. https://autlawiel.files.wordpress.com/2014/07/af-kiribati-2013-nzipt-800413.pdf. Accessed February 2018.

[12] Convention relating to the Status of Refugees, adopted 28 July 1951, entered into force 22 April 1954, to be read in conjunction with the Protocol Relating to the Status of Refugees, adopted on 31 January 1957.

will be able to manage this specific situation, so that the interested individuals will be granted some protection in a dignified manner. Today's migrant however does not enjoy a similar possibility.

This represents a significant complication in the study of the issue, and makes the calculation of those who will be involved in the future extremely difficult, if not *tout court* impossible. An individual can migrate because of the expectation that his or her home will become uninhabitable because of climate change and environmental degradation, and this movement remains to a certain extent at present voluntary: it is simply a form of voluntary adaptation to gradual environmental changes driven by global warming. By contrast, a person could be obliged to leave his or her home and emigrate from the habitual residence (or country) when every other viable option is no longer available: if there is no chance to live further in a specific region because this has become uninhabitable due to climate change, then the migrant becomes a real "climate refugee[13]". This distinction has been observed, for example, in studies related to the responses to the different stages of a famine.[14]

In all this, Renaud's classification of "environmental migrants" is useful. He uses the degree of "forcedness" of the migration as a distinctive criterion, by this means capturing in a comprehensive manner the various dimensions of displacement that can be associated with environmental degradation and global warming.[15]

i. "Environmentally motivated migrants" (EMMs) are individuals who decide to migrate preventively in order to distance themselves from future and significant environmental degradation. Their decision to migrate is largely dictated by already existing socio-economic difficulties, but also by their expected worsening due to

[13] Cfr. Hugo (2010). Climate Change-Induced Mobility and Existing Migration Regime in Asia and the Pacific. In McAdam (Ed.) (2010) (op. cit.), p. 13.

[14] Bohle, Cannon, Hugo & Ibrahim (Eds.) (1991). *Famine and Food Security in Africa and Asia: Indigenous Response and External Intervention to Avoid Hunger*. Bayreuth: Bayreuther Geowissenschaftilche Arbeiten.

[15] Renaud, Bogardi, Dun & Warner (2007). Control, Adapt or Flee: How to Face Environmental Migration? *InterSecTions Publication Series*, 5. UNU-EHS. https://collections.unu.edu/view/UNU:1859#:~:text=Control%2C%20adapt%20or%20flee%3A%20how%20to%20face%20environmental%20migration%3F,-Series%20Title&text=The%20complex%20subject%20of%20this,sudden%20onset%20of%20hazard%20events. Accessed 10 June 2019.

climate/environmental events and processes. At the time of the migration, EMMs do not live in a situation of current or imminent danger and environmental push-factors play merely *a* role together with others. These people are and will therefore be considered as mere economic migrants who leave their countries in search of better economic opportunities. Thus, given the (albeit limited) possibility of remaining in their own country and given the lack of a condition of extreme difficulty, it is unlikely that significant international protection regimes could be invoked for them.

ii. "Environmentally forced migrants" (EFMs) face situations in which some environmental changes are destroying or are about to destroy their livelihoods, and therefore cannot choose to stay and avoid to migrate, even if a certain margin of choice regarding "when" to leave remains.

iii. "Environmental emergency migrants" (EEMs) are people with no choice about both whether and when to migrate. EEMs are forced to abandon their places of origin due to extremely serious climate or environmental impacts which completely make the "staying option" not available. Similar hypotheses can occur due to natural disasters (both sudden and gradual) capable of threatening the very survival of the affected individuals.

In general, alongside the complex causal relationship and the difficulty of distinguishing between forced and voluntary migration, there is a third fundamental aspect for the purposes of a correct conceptualization of the phenomenon of "climate refugees": many of the people affected by environmental changes—including climate-driven ones—do not have the means and the possibilities to emigrate. Emigrating presupposes a certain financial availability together with a reliable migration network of acquaintances already existing abroad.[16] Unluckily in this regard, the countries that are most vulnerable to generating "climate refugees" are

[16] Migration networks appear when a pioneer migrant reaches a destination and begins to settle there. Over time, the pioneer will be able to provide his/her family and friends with reliable training on new life, job opportunities, assistance on arrival, etc. This theory explains clearly how and why emigration sites are established and maintained in the long run, as it explains why migrants tend to concentrate in certain places, rather than randomly choosing others. This theory is also the basis of the climate change adaptation project promoted by the Kiribati government.

often very poor developing countries (see those in the Sahel region or Bangladesh, for example). Many of the individuals affected do not enjoy sufficient resources to undertake an international journey and start a new life abroad. They will more probably be forced to stay in disastrous environmental conditions, or at best become internally displaced persons.

With this, we reach a further pivotal point: despite public discourse, the vast majority of "climate refugees" will not cross national borders, moving instead within their own nations. In fact, empirical research on migration shows that in most cases, migration remains internal to any given state.[17] This however contrasts with part of the maximalist literature and of the media coverage accorded to it, which tend to consider all climate-induced migrants as international ones. Actually, this is not the case.

In fact, already today some isolated populations are obliged to relocate internally due to climate change-linked processes. It is worth mentioning in this regard some Alaskan villages populated by Native Americans who, due to soil erosion and floods brought about by rising sea levels, have chosen to leave their territories and migrate elsewhere on US territory.[18] A similar fate affects the inhabitants of the Island of Jean Charles, in Louisiana, which has lost up to 98% of its land due to sea-level rise and irresponsible fossil fuel extraction in the area. Recently, in order to remedy the situation, the US federal government has allocated financial resources with the aim of relocating the inhabitants of the island, who belong to an ancient tribe of Native Americans that for over one hundred and seventy years have been living there.[19] Similarly, in the last decades,

[17] For a detailed study undergone in twenty-three different countries, see EACH-FOR (2009). Environmental Change and Forced Migration Scenarios, Synthesis Report. http://rosamartinez.org/wp-content/uploads/2015/11/Migraciones-y-Cambio-Climatico_EACHFOR.pdf. Accessed 2 September 2018. See also: Laczko & Agharzam (Eds.) (2009). *Migration, Environment and Climate Change: Assessing the Evidence*. Geneva: IOM.

[18] See in this regard US Government Accountability Office (2009). Alaska Native Villages: Limited Progress Has Been Made on Relocating Villages Threatened by Flooding and Erosion. Report to Congressional Requesters. https://www.gao.gov/assets/300/290468.pdf. Accessed 15 May 2019.

[19] D'Angelo (2016, December 2). A Louisiana Tribe Is Now Officially a Community of Climate Refugees. *Huffington Post Hawaii*. https://www.huffpost.com/entry/climate-refugees-louisiana_n_56bbd5efe4b0c3c550501784?guccounter=1&guce_referrer=aHR0cHM6Ly93d3cuZ29vZ2xlLmNvbS8&guce_referrer_sig=AQAAAAcHPSHZKmfQvaam-92okhuewL22cmI0xVjAooX_VgJLoxMLeMitEZ8vx9OpvK-dqLh3DvXH6VQ6ojgsvgXl0MhrbkMWCxjTMXDXzyFM18gDaDQnpWk43ZDTMkPjgCK61_7SELfa-dTMlVe5962oletUa7IvGr9kf_XI85s8xBiu. Accessed 14 April 2019.

five (uninhabited) islands of the Solomon Islands, in the Pacific Ocean, have been completely submerged, while other inhabited ones are gradually increasingly invaded by the sea, with at least two villages that have effectively been "erased" by growing tides and coastal erosion.[20] Lastly, similar events affect the inhabitants of the Carteret Islands, in Papua New Guinea. These populations, through an internal migration/relocation plan called "*Tulele Peisa*" (Sailing the Waves on our Own), have decided to leave their origin islands and move to other areas of their country.[21] Thus, although using an improper legal terminology, the populations of these small villages across different countries in the world can be considered as the first (internal) "climate refugees" in history. Though affecting relatively small numbers of people so far, these relocation programs are far from being dealt with easily, stuck between the insufficient allocation of funds and the inherent difficulty of the inhabitants to leave their ancestral homes. Cast in a broader light, Sumudu Atapattu makes her point when she writes that "if the wealthiest country in the world cannot address the plight of less than 1,000 citizens, it is hard to imagine how the international community will cope with the plight of a large number of people who could be forced to move as a result of climate change[22]".

The data on the importance of internal displacement, which on average more than doubles the number of international refugees,[23] are important for practical and legal purposes, too. If the vast majority of climate-induced migrations is internal, it is opportune to critically consider the excessive focus, with a view to protecting the individuals involved, on international protection mechanisms derived from the model of the 1951

[20] Albert et al. (2016). Interactions Between Sea-Level Rise and Wave Exposure on Reef Island Dynamics in the Solomon Islands. *Environmental Research Letters*, 11, p. 1.

[21] See the dedicated website: www.tulele-peisa.org. Accessed 1 January 2020. See on this matter: Pascoe (2015). Sailing the Waves on our Own: Climate Migration, Self-Determination and the Carteret Islands. *QUT Law Review*, 2, p. 72 ff.

[22] Atapattu (2018). A New Category of Refugees? "Climate Refugees" and Gaping Hole in International Law. In Behrman & Kent (Eds.), *'Climate refugees' Beyond the Legal Impasse?* New York: Routledge, p. 35.

[23] See the data collected since 1990 by the Internal Displacement Monitoring Center. Especially with the beginning of 2000, the number of IDPs has constantly more than doubled that of refugees each year. See http://www.internal-displacement.org/database/displacement-data, table on Number of IDPs and Refugees. Accessed 7 February 2020.

Refugee Convention. In fact, these instruments only apply to cross-border migrations. We must therefore ask ourselves what could and should be the role of international law in the management and protection of internal migrants, since this phenomenon plays out primarily within national borders. The Guiding Principles on Internal Displacement (GPID), adopted by the UN General Assembly in 1998, are certainly helpful in this regard, providing the international community and states that want to implement them internally with a remarkable reference model. It is clear that different legal frameworks apply to internal and international climate migrants, and that it is crucial to understand that some line has to be traced within the "climate refugees cauldron" if one wants to provide viable (legal) solutions for the management of this emerging topic.

Therefore, when dealing with "climate refugees", it is important to carefully consider what research has highlighted so far, and, namely, that:

i. climate change represents in most climate-related migrations "only" a threat multiplier,
ii. the distinction between voluntary and forced climate migrants may be ontologically blurred, but is of utmost importance legally speaking,
iii. the vast majority of so-called climate refugees stays within its own country, without crossing international borders.

These elements are to be seriously taken into account, lest the risk of directing policy-makers and legislators toward a rough and incorrect understanding of the problem. Quite often, the issue of "climate refugees" is described in the media and political discourse as a global, massive, and international challenge, where global warming is depicted as the sole and fundamental cause of an emerging refugee crisis. This standpoint makes sense only from an extremely general perspective, that does not account for all those elements and considerations we just underlined with regard to the high degree of complexity that lies behind this phenomenon. Understandably, media and political discourse lean and have leant toward such interpretations because of an unavoidable demand for sensationalism by the market. Interestingly, however, much legal and political research too has backed these interpretations, at the point that Benoit Mayer, one of the most enlightening scholars in the field of "conceptualizing climate refugees", has wondered "what made it possible for the law and policy literature to get its analytical premises so wrong for such a long time[24]". In the following pages, we will also focus on this matter, delving in the analysis of the two main schools of thoughts that have studied "climate refugees" over the years: maximalism and minimalism.

In conclusion, many estimates regarding the number of those who are and will be "climate refugees" are based on questionable research methodologies: when studies do not sufficiently take into consideration the multi-causality of migration choices, the difficulty in distinguishing forced and voluntary migration, and the fact that most of the affected individuals move within their own countries, they become ultimately unreliable. The IPCC itself, while stating, as we already saw, since 1990 that the single most challenging effect of global warming would be on human mobility, has also been constantly repeating that there is little confidence in quantitative forecasts regarding changes in the mobility of populations, and this is due to their complex and multi-causal nature.[25]

[24] Mayer (2018). Who Are "Climate Refugees"? Academic Engagement in the Post-truth Era. In Behrman & Kent (Eds.) (op. cit.), p. 90.

[25] IPCC (2014). Summary for policymakers. In *Climate Change 2014: Impacts, Adaptation, and Vulnerability. Part A: Global and Sectoral Aspects*. Cambridge: Cambridge University Press, p. 20. https://www.ipcc.ch/site/assets/uploads/2018/02/WGIIAR5-PartA_FINAL.pdf. Accessed 15 August 2019.

CHAPTER 4

How Many "Climate Refugees"? Pros and Cons of Maximalism and Minimalism

There are very different estimates regarding the number of people who migrate or will migrate due to climate change. The academic debate on these differences is one of the most remarkable aspects when we study our topic.

The UN first began to deal with it in the middle of the Eighties, when Essam El-Hinnawi used the term "environmental refugee" in the report of the United Nations Environment Programme (UNEP) to highlight the potential impacts of uncontrolled development and pollution on migration flows on a global scale.[1] Since then, research has proliferated. However, given the inherently complex (and often unpredictable) nature of the two fields underpinning the phenomenon—climate patterns and human mobility—separate discourses, approaches, and research methods have developed over the years. The study of so-called climate refugees has thus divided itself into two main schools of thought: the maximalist and the minimalist ones. The two differ greatly with regard to their predictions relating to the number of those who are (and will be) forced to abandon their places of origin due to climatic/environmental alterations. This chapter critically analyzes the debate that has arisen between maximalist and minimalist scholars, highlighting the pros and cons of both

[1] El-Hinnawi (1985). Environmental Refugees. Report. UNEP. https://digitallibrary.un.org/record/121267. Accessed 12 February 2020.

G. Sciaccaluga, *International Law and the Protection of "Climate Refugees"*,
https://doi.org/10.1007/978-3-030-52402-9_4

schools. It starts looking at the maximalist approach and its critiques, moving on to discussing the minimalist thought. In the end, this chapter argues that the two should not be viewed necessarily as incompatible, but rather as two schools that study two different dimensions (or layers) of the same phenomenon.

Since 2008, an annual average of 27 million individuals has been forcibly removed from their homes due to natural disasters, be they of meteorological, geophysical, or climatic origin.[2] As Walter Kälin[3] pointed out vividly, this means one person per second.[4] From 1970, the probability of being displaced due to natural disasters has doubled because of both increased demographic pressure and the effects of global warming, which increases the frequency and intensity of extreme meteorological events. Together with the certainty that climate change will keep worsening whatever the mitigation and adaptation measures undertaken internationally, these figures suggest that the number of individuals involved in migrations caused by climate or environmental pressures is destined to increase in the near future. Nevertheless, whereas it is certain that the numbers will increase, the exact amount remains in the realm of unknown things. Relying on current scientific knowledge, it is complex (or, better, impossible) to assert with certainty how many people are or will be forced to migrate because of the negative effects of climate change, as well as to predict exactly when and where this will happen. However, when the IPCC, beginning in the 1990s, declared that migration would be one of the main effects of climate change,[5] various authors began to attribute a direct causal relationship between global warming and migratory flows.

This was the beginning of what was later labeled the maximalist school of thought. According to this orientation (in truth today quite obsolete

[2] The IDMC is the most authoritative non-governmental organization within the international community with regard to the monitoring of internal (but also external) human mobility. Established in 1998, it is supported and recognized by the UN General Assembly. For more information, see the website www.internal-displacement.org.

[3] Formerly Representative of the United Nations Secretary-General on the human rights of internally displaced persons and one of the greatest experts in the world on the subject.

[4] See IDMC (2015). People Displaced by Disasters. Global estimates 2015. Report. IDMC. https://www.internal-displacement.org/sites/default/files/inline-files/20150713-global-estimates-2015-en-v1.pdf. Accessed 18 March 2020.

[5] See IPCC (1990). First Assessment Report. Climate Change: The IPCC Impacts Assessment (1990), Policy-Makers' Summary. https://www.ipcc.ch/site/assets/uploads/2018/05/ipcc_90_92_assessments_far_full_report.pdf. Accessed 15 March 2020, p. 3.

within academic circles but still in vogue in the media), hundreds of millions of people, even up to a billion, will be forced to migrate due to climate change within our century.[6] Maximalists (sometimes also referred to as alarmists) place the issue of climate-induced migration in the broader discourse aimed at highlighting the dangers arising from climate change. In this view, they emphasize the extremely large numbers of people forced to migrate because of global warming as a way of showing its destructiveness. By so doing, the alarmist school has served the purpose of raising not only academic but also political attention to the importance of climate change mitigation policies.[7]

Despite the benefits of greater climate awareness, the alarmist school presents however some obvious methodological problems, since it tends to ignore the results of empirical studies on migration, providing and proposing a rather simplistic reading of the problem. The most well-known maximalist author, Norman Myers, an Oxford University scholar in environmental sciences and still one of the most cited academics in the field of climate-related migration, has, for instance, been strongly

[6] See, for instance: Christian Aid (2007). Human Tide: The Real Migration Crisis. https://www.christianaid.org.uk/sites/default/files/2017-08/human-tide-the-real-migration-crisis-may-2007.pdf. Accessed 13 April 2019. Friends of Earth in Australia (2007). A Citizen's Guide to Climate Refugees. https://www.safecom.org.au/pdfs/foe-climate-citizens-guide.pdf. Accessed 12 May 2019. McKie (2010, 28 November). Climate Change Will Cost a Billion People Their Homes, says report. *The Observer* https://www.theguardian.com/environment/2010/nov/28/cancun-climate-summit-weather. Accessed 13 May 2019. International Federation of the Red Cross and Red Crescent Societies (2001). World Disasters Report: Focus on Recovery. Report. IFRC and Red Crescent Societies. Myers (1993). Environmental Refugees in a Globally Warmed World. *Bioscience*, 43, p. 752. Myers (2005). Environmental Refugees: An Emergent Security Issue, 13. Economic Forum, Prague, 23–27 May 2005. https://www.osce.org/eea/14851?download=true. Accessed 12 May 2018.

[7] In this sense, see Bettini (2013). Climate Barbarians at the Gate? A Critique of Apocalyptic Narratives on 'Climate Refugees'. *Geoforum*, 45, p. 65. https://www.sciencedirect.com/science/article/abs/pii/S0016718512001947?via%3Dihub. Accessed 12 January 2020: "Moreover, a collateral to apocalyptic narration is the envisioning of an ultimate threat (the tsunamis of climate refugees), and raises the imperative 'If we do not do this... some unspeakably horrible will take place'. It thereby mobilizes fear and a sense of urgency". See also Mayer (2012). 'Environmental Refugees'? A Critical Perspective on the Normative Discourse. Center for Sustainable Development Law. https://papers.ssrn.com/sol3/papers.cfm?abstract_id=2111825. Accessed 14 January 2016, p. 2: "The limpid simplicity of the maximalist estimations and previsions exerts a certain attraction, providing a simplistic, media-friendly, and policy relevant message [...] – a discourse more efficient than the halftone, prudent voices of minimalists".

criticized for his research methodology. Back in 1993 for example, he wrote that by 2050 there would be 150 million people displaced due to climate change. To obtain these figures, he identified the parts of the globe that are most vulnerable to phenomena such as sea-level rise and extreme weather events. On that basis, relying on the expected rates of population growth in those areas for the following decades, he considered the number of people possibly subject to forced migration. Myers subsequently reviewed these numbers, going so far as to assume that they would reach 250 million by 2050. The problem with his methodology emerged quite soon: he considered as "environmental refugees" and therefore, technically, as international forced migrants, almost every individual deciding to migrate (even internally) for reasons that, more or less directly, were linked to environmental and climate factors.[8] In this perspective, anyone deciding to leave his or her country *also but not only because* of droughts, floods, or famines, is immediately thrown into the environmental refugees "cauldron" without any adequate analysis—at least from a legal point of view—regarding the relevant distinctions to be made within this "cauldron".

Highlighting the lack of precision of this method is a 2005 publication, which reads: "to the extent that this article postulates that the highest concentrations of environmental refugees are located in developing regions, OECD countries may respond that the problem 'down there' has little to do with them in practical terms. [...]. Developed countries cannot [however] isolate themselves from the emergencies and disasters of developing countries: there are considerable numbers of environmental refugees who have already entered, usually illegally, in OECD countries. [...]. Consider, for example, the case of Italy, a country well positioned for North Africans, which can clandestinely navigate from Tunisia to Sicily in three hours. Their number is, at least, 120.00 a year and their total has exceeded one million [...]".[9] Now, although many migrants that come to Italy from various African states certainly suffer also

[8] "Myers' methodology was rudimentary, boing down to more little more than adding up the populations living within exposed areas". In Mayer (2018). Who Are "Climate Refugees"? Academic Engagement in the Post-Truth Era. In Behrman & Kent (Eds.) (2018). *'Climate refugees' Beyond the Legal Impasse?* New York: Routledge, p. 92.

[9] Myers (2005). Environmental Refugees: An Emergent Security Issue, 13. Economic Forum, Prague, 23–27 May 2005. https://www.osce.org/eea/14851?download=true. Accessed 12 May 2018, pp. 3–4.

from climatic or environmental pressures, it is exaggerated and incorrect to consider them as pure "environmental refugees". These migrants are also pushed toward Italy and Europe more broadly by armed conflicts, poverty, governance failures, political persecutions, and so on. Viewed from this perspective, the maximalist approach is highly inadequate, since it decidedly oversimplifies environmental-related migration phenomena, up to the point of "betraying the mission of academia to provide the best information possible as the basis for public policy".[10]

Despite this clear methodological weakness, however, the maximalist approach is not to be discarded completely. In fact, maximalist studies still make a strong case if cast in a broad framework. A relatively recent study published by the US National Academy of Sciences has shown, for example, that in the outbreak of the Syrian civil conflict, climate change played a decisive role. In the years preceding the conflict, from 2007 to 2010, the country experienced one of the harshest droughts recorded in its history. According to the authors of the paper, global warming exacerbated to a certain but indeterminable degree the drought, which, in turn, led many Syrian farmers to look for alternative economic opportunities in urban areas, exacerbating the social pressures that weighed in on an already unstable political context. This led to a series of social, economic, and political tensions, which in the end brought to the outbreak of the conflict.[11] This study enjoyed considerable media resonance,[12] to the

[10] Mayer (2018). In Behrman & Kent (Eds.) (op. cit.), p. 91.

[11] Kelley et al. (2015). Climate Change in the Fertile Crescent and Implications of the Recent Syrian Drought. *Proceedings of the National Academy of Science*, 11, pp. 3241 ff.

[12] See Welch (2015, March 2), Climate Change Helped Spark Syrian War, Study Says. *National Geographic*. https://www.nationalgeographic.com/news/2015/3/150302-syria-war-climate-change-drought/. Accessed 10 June 2019. Bawden (2015, March 2), Climate Change Key in Syrian Conflict—And It Will Trigger More War in Future. *The Independent*. https://www.independent.co.uk/news/world/middle-east/climate-change-key-in-syrian-conflict-and-it-will-trigger-more-war-in-future-10081163.html. Accessed 10 June 2019. Fountain (2015, March 2). Researchers Link Syrian Conflict to a Drought Made Worse by Climate Change. *The New York Times*. https://www.nytimes.com/2015/03/03/science/earth/study-links-syria-conflict-to-drought-caused-by-climate-change.html. Accessed 9 June 2019. See also Wendle (2015). The Ominous Story of Syria's Climate Refugees. *Scientific American*. https://www.scientificamerican.com/article/ominous-story-of-syria-climate-refugees/. Accessed 14 April 2019. Randall (2016). Syria and Climate Change: Did the Media Get It Right? *Climate and Migration Coalition*. http://climatemigration.org.uk/syria-and-climate-change-did-the-media-get-it-right-2/. Accessed 19 May 2019.

point of inducing then US Secretary of State Hillary Clinton's staff, toward the end of 2015, to recognize that climate change is to be considered as one of the causes of the Syrian civil war and, more generally, as one of the underlying factors behind the forced displacement of entire populations.[13] In short, climate change was considered as a trigger of the Syrian civil war and the refugee crisis that followed. As further proof of the correlation and possible causation between global warming and uncontrolled migration events, one can also look at the exodus from Somalia to Kenya between 2012 and 2013, driven both by a severe drought and by the violence perpetrated by the Al Shabab regime.[14]

If this is valid, anthropogenic climate change can be viewed as already contributing—though in a complex and multi-causal way—to massive migration phenomena. If we consider that even in the event of successful global warming mitigation policies, the increase in the average global temperatures will double compared to the recent past, it becomes quite clear that the international community will face intense challenges associated with climate change-related migration pressures. If the maximalist research sins, therefore, of an excessive simplification of reality, that transforms climate change and environmental degradation into a phagocytizing push-factor for migration, it nonetheless can clearly show that global warming already plays a decisive role in triggering massive migratory events.

On the one hand, authors like Myers have had the undoubted merit of bringing the topic to the attention of a broad public; on the other,

[13] Merica (2015, November 4). Hillary Clinton: Climate Change Has Contributed to Refugee Crises, Including Syria. *CNN*. https://edition.cnn.com/2015/11/04/politics/hillary-clinton-climate-change-syria-refugee-crises/index.html. Accessed 13 March 2019. See also the 2015 declarations by the former US Secretary of State USA John Kerry: "I'm not saying that the crisis in Syria was created by climate change. But the devastating drought made a bad situation a lot worse", in Brown (2016). Europe's Refugee Crisis: A Taste of Things to Come? *UNEP's Disaster and Conflicts Subprogramme*. https://medium.com/@UNEP/europe-s-refugee-crisis-a-taste-of-things-to-come-72a5d4079cb4. Accessed 30 May 2019.

[14] UNHCR (2015). UNHCR, the Environment & Climate Change (Updated Version). Report. UNHCR. https://www.unhcr.org/540854f49.pdf. Accessed 14 June 2019, p. 5. See also Afifi, Govil, Sakdapolrak, & Warner (2012). Climate Change, Vulnerability and Human Mobility: Perspectives of Refugees from the East and Horn Africa. Report. UNHCR & UNU-EHS, https://www.unhcr.org/protection/environment/4fe8538d9/climate-change-vulnerability-human-mobility-perspectives-refugees-east.html. Accessed 14 June 2019.

their often "explosive" predictions, combined with the use of a legally improper terminology and a problematic methodology, have contributed to creating an ill-informed debate in which the most alarmist and sensationalist authors obtain the greatest media and public attention. Mayer brilliantly explains this situation: "certainly, the limpid simplicity of the maximalist estimations and previsions exerts a certain attraction, providing a simplistic, media-friendly and policy-relevant message [...] - a discourse more efficient than the halftone, prudent voices of the minimalists".[15]

"*Un miliardo di rifugiati climatici entro il 2050*" (A billion climate refugees by 2050), is, for instance, the title chosen by the moderate Italian newspaper "La Stampa" in 2017 to highlight the problem of climate-related displacement.[16] Relying on the 2017 report of the Lancet Countdown, "La Stampa" reported these exorbitant number—associated with misuse of legal terms—following the far more sober (but still alarmist) wording chosen by the Lancet Countdown Report, which sound as follows: "the total number of people vulnerable to migration might increase to 1 billion by the end of the century without significant further action on climate change".[17] In the same vein, Jane McAdam pointed out, for example, that in the days preceding the 2010 Cancun Conference on climate change (UNFCCC COP 16), an article published by "The Observer" talked in dramatic terms of a "special report" which predicted that up to one billion people would have lost their homes over the next ninety years due climate change.[18] The reported study, however, was based on dramatic global warming scenarios: an increase of 4°C or more in 2100 compared to 1850. Furthermore, the report by François Gemenne, another leading author in the study of climate-driven migrations, did not state at all that one billion people would be forced to

[15] Mayer (2012). (op. cit.), p. 3.

[16] Giovannini (2017, November 6). Un miliardo di rifugiati climatici entro il 2050. *La Stampa*. https://www.lastampa.it/tuttogreen/2017/11/06/news/un-miliardo-di-rifugiati-climatici-entro-il-2050-1.34380312. Accessed 13 January 2020.

[17] (2017). The Lancet Countdown on Health and Climate Change: From 25 Years of Inaction to a Global Transformation for Public Health. https://www.thelancet.com/journals/lancet/article/PIIS0140-6736(17)32464-9/fulltext. Accessed 13 May 2019.

[18] McKie (2010). (op. cit.).

leave their homes due to climate change, but "only" that they could be negatively affected by water availability alterations.[19]

Ultimately, the figures put forward by what we may call "media-savvy maximalism"[20] are quite unreliable from a scientific point of view. They do not sufficiently take into account the multi-causality and multi-dimensionality existing in any choice to migrate, just as they do not consider the fact that many individuals affected by climate change may decide to adapt *in situ* or, more simply, that they might opt to migrate within their own country. We should emphasize, for example, that even in the context of "drowning states" where sea-level rise has impacted on more than 90% of the households, only 8–23% of migrants (both internal and/or international) identifies climate change as one of their decisive push-factors.[21] This does not come as a surprise when one considers the socio-economic conditions of a state like Kiribati, for example, where demographic pressure and economic weakness have been present for decades. At the same time, however, it is equally obvious that in similar arduous contexts, the worsening of environmental conditions caused by global warming could become "the straw that breaks the camel's back". More than 70% of I-Kiribati and Tuvaluan households have stated that, over time, their migration option will be increasingly determined by drought and sea-level rise.[22]

In general, one of the greatest difficulties in understanding the link between migration and climate is due to the unpredictability of human adaptability and, consequently, to the degree of resilience of an individual or a community in facing environmental pressures. Stating that the inhabitants of the coastal areas of the globe subject to rising oceans *will be forced* to migrate is simply wrong. Despite all the problems that these individuals face, they have the option of adapting and staying where

[19] Gemenne (2011). Climate-Induced Displacement in a 4°C+ Degree World. *Philosophical Transactions of the Royal Society A*, 182.

[20] See also Doherrty & Slezak (2017, August 10). Australia Faces Potentially Disastrous Consequences of Climate Change, Inquiry Told. *The Guardian*. https://www.theguardian.com/environment/2017/aug/11/australia-potentially-disastrous-consequences-of-climate-change-inquiry-told. Accessed 12 January 2019.

[21] United Nations University (Institute for Environment and Human Security) (2014). Climate Change and Migration in The Pacific: Links, Attitudes, and Future Scenarios in Nauru, Tuvalu, and Kiribati, Fiji. Report. UN University. https://collections.unu.edu/view/UNU:6515. Accessed 14 June 2019.

[22] Ibid.

they are, to migrate within their own country, or to migrate for reasons that could be, for a given individual, more pressing than climatic ones. Most of the studies foreseeing large numbers of "climate refugees" are based on an implicit climate risk assessment, rather than on a careful and objective study of the relationship between migration and climatic factors. Despite these methodological shortcomings, the figures suggested by maximalism are often used in the scientific literature and by several international organizations as well. The "Stern Report", for example, an authoritative and highly influential macroeconomic document that studied the costs and benefits of climate change and its corrective policy, described Myers' figures as conservative.[23] In 2008, the International Organization for Migration (IOM) estimated that by 2050 there would be 200 million "environmental refugees",[24] thus positioning itself halfway between Myers' first predictions of 25–50 million and the most "explosive" maximalist ones that talk of 700 million individuals forced to migrate due to environmental pressures.[25]

Of course, the fact remains that we still need to see how much the Earth will actually warm up due to the greenhouse effect, but in the vast majority of cases, these enormous figures are reported by the media without questioning the methodology used and, ultimately, accounting for the complexity of a phenomenon that is too often oversimplified. The implicit political intent behind these studies to galvanize attention to ensure better protection of the environment and climate stability in the long run is appreciable. From an ethical point of view, by making today's generations more responsible for future ones, maximalists play a positive and important role. Thanks to their efforts, large numbers of people all over the world are becoming acquainted with the theme of environmental/climate displacement, feeding also reflections on the broader issue of the relationship between humankind and the environment it lives and develops in.

Next to this virtue and alongside its methodological shortcomings, the maximalist approach has questionable political effects too. "Trumpeting

[23] Stern (2007). *The Economics of Climate Change: The Stern Review*. Cambridge: Cambridge University Press, p. 112.

[24] Brown & Oli (2008). Migration and Climate Change. International Organization for Migration: Research Series No. 31. https://publications.iom.int/system/files/pdf/migration_and_environment.pdf. Accessed 13 May 2019.

[25] Christian Aid (2007). (op. cit.).

such figures in the absence of defensible methodologies can have a deleterious effect. When the empirical evidence does not support them, the very motivation behind the maximalist approach - engendering action - is undercut, because people perceive that there is no clear evidence".[26] All this tends to fuel an ill-informed and polarized debate on climatic and environmental issues. A further negative effect of an alarmist approach is that of fearmongering about potential migratory movements, which, according to some, would become the cause of conflicts, cultural loss, and epidemics.[27] As a matter of fact, uncontrolled (climate-induced) migrations can represent one of the multiple causes of armed conflicts in already unstable political situations. Recent studies show that the risk of conflict in areas characterized by the presence of different ethnic groups increases in the presence of disasters and phenomena related to anthropogenic climate changes.[28] Similar conclusions were also reached by the UN Security Council in 2011, when it expressed concern about the role that climate change could have, in the long run, over the exacerbation or outbreak of armed conflicts.[29] This led in 2017 to the Security Council's explicit recognition[30] that there is a clear correlation between the effects of climate change and regional and national stability.[31] In this regard, the discoveries regarding the relationship between global warming, drought, and the Syrian civil conflict tend to confirm the theory that climate

[26] McAdam (2012). (op. cit.), p. 29.

[27] See for instance: Homer-Dixon & Percival (1996). *Environmental Scarcity and Violent Conflict: Briefing Book*. Toronto: University of Toronto and the American Association for the Advancement of Science. According to the authors, migrations forced by climate change would generate conflicts as a result of: an increase in competition for increasingly scarce natural and economic resources; a consequent increase in existing ethnic/cultural tensions; greater demand for social infrastructure and, therefore, political discontent.

[28] Cf. Schleussner et al. (2016). Armed-Conflict Risks Enhanced by Climate-Related Disasters in Ethnically Fractionalized Countries. *PNAS*, 33. https://www.pnas.org/content/113/33/9216. Accessed 14 October 2019.

[29] UN Security Council Statement, in occasion of the 6587th Meeting, 20 July 2011 (UN Doc. S/PRST/2011/15).

[30] UN Security Council resolution 2349 (2017) [on the situation in the Lake Chad Basin region] S/RES/2349; UN Security Council Resolution 2408 (2018) S/RES/2408 (2018).

[31] Motzfeldt Kravik (2018). The Security Council and Climate Change—Too Hot to Handle? *EJIL: Talk*. https://www.ejiltalk.org/the-security-council-and-climate-change-too-hot-to-handle/. Accessed 14 September 2019.

change, whatever its manifestation, contributes to the outbreak or to the worsening of conflicts in fragile socio-economic and geopolitical situations.

Excessively alarmist predictions fuel a widespread fear of future uncontrollable migration flows, often viewed and used to exacerbate fears related to loss of identity, economic opportunity, and security. In this way, migrations tend to be considered as intrinsically negative events and, consequently, as something that has to be stopped, leading to a dangerous and acritical approach that denies the existence of any positive effect linked to human migration.[32] It is not among the aims of this book to investigate the advisability, or otherwise, of closure policies in reaction to (climate) migrations. Nevertheless, a brief discussion is useful for the purposes of this study.

When the issue of "climate refugees" is looked at through alarmist lenses, it inevitably tends to be framed as a security question rather than a humanitarian one.[33] Excessively alarmist and wrong predictions contribute, in fact, to creating fear of excessive and uncontrollable movements, crediting the theses maintaining that states, in order to manage the phenomenon, must close their borders to stop migratory flows. A possible complementary consequence of these policies would be to try to box in climate migrants within the affected countries, notoriously among the poorest in the world.[34] By so doing, the basic principles of international cooperation and humanity would be put in question. It is possible, if not probable, that such a "remedy" would exacerbate the problem, condemning countless people to severe suffering and preventing

[32] Bakewell (2008). Keeping Them in Their Place: The Ambivalent Relationship between Development and Migration in Africa. *Third World Quarterly*, 7 (29), pp. 1341 ff.

[33] Jastram (2014). Warm World, Cold Reception: Climate Change, National Security and Forced Migration. *Vermont. Journal of Envtl. Law*, 2014, pp. 752 ff.

[34] It is well known that the countries most affected by climate change-related migrations will be among the poorest in the world (for example, Bangladesh, Sahel states, micro-state islands, etc.) or, in any case, among those least able to respond to environmental degradation (India, Pakistan, China, Egypt, Vietnam, etc.). In summary, the factors that contribute to increasing vulnerability to the negative effects of climate change are poor education, weak institutions, scarce availability of resources, poor technical and technological skills, and poverty.

the development of international cooperation on migration.[35] A governance of migration, at national, regional as well as international level, should try to channel these policies toward conditions of security and legality, instead of seeking to prevent them *tout court*.[36] Considering potential climate migrations as a threat to developed countries exacerbates the North–South dichotomy, fueling a *Weltanschauung* of one against the other and preventing a global and interdependent humanitarian and environmental protection system.

One of the greatest contemporary Western thinkers, Tzvetan Todorov, has wonderfully shed light on the absurd and even anti-historical effects of excessive cultural containment. Although his thesis refers to the broader theme of multiculturalism and the genesis and evolution of collective identities, his ideas can be taken up to avoid deleterious policies on migration and climate change. "Every society is multicultural. [...]. The careful management of this growing plurality implies not to assimilate others to the majority culture, but to respect minorities, integrating them into a framework of laws and civic values common to all. [...]. In fact, the separation and closure of cultures or communities are closer to the pole of barbarism, while their mutual recognition is a step towards civilization. Public money should preferably go to what brings together, not what isolates".[37] Obviously, an approach open to the establishment of international, top-down migration policies can be followed more easily in certain geopolitical and geographical contexts, less in other ones. It is possible to assume, for example, that over the years, the forced migration from micro islands in the Pacific Ocean will be managed in a dignified manner. This because of the relatively small amount of people involved (ca. one million), of the geographical characteristics of the nations involved, and because the most probable destination states (such as Australia, New Zealand, and the United States of America) are wealthy and prosperous

[35] Notwithstanding the fact that policies aimed at stopping *tout court* migrations movement rarely work.

[36] Gemenne insists on the need to promote the right to mobility and, to this end, to replace the "security agenda" with a "mobility agenda". In other words, he proposes to focus the debate on the "climate refugees" on the development and implementation of effective and safe migration policies, devoting energy to the creation of migratory corridors, as well as to safeguarding the cultural values of the emigration communities and the destination ones.

[37] Todorov (2009). *La paura dei barbari: oltre lo scontro delle civiltà*. Milan: Garzanti, p. 268.

countries. Not surprisingly, the World Bank has suggested Australia and New Zealand to create legal immigration pathways from low-lying micro-states so as to avoid mass migrations in the future.[38] A similar approach however is less likely to hold in other geographical contexts. The delta of the Ganges River, where Bangladesh is located, is one of the most climatically vulnerable zones in the world. This particular geographical condition, in a context of ascertained poverty, high population density, and institutional fragility, is likely to lead to substantial migratory movements, especially toward India, a country that is geographically and culturally close to Bangladesh. Also in anticipation of these flows, the governments of the two countries decided in 2005 to build a barbed-wire wall along broad tracts of their border with the aim of preventing and impeding excessive irregular migrations, anticipating Donald Trump's desired wall along the United States–Mexico border.

Fully aware of the complexity of the challenge that lies ahead and also of the difficulty of adopting alternative solutions for the management of migratory flows in a system of sovereign states, this author argues that the climate migrants phenomenon should be conceived in terms of safeguarding the people affected and not (only) in terms of the survival of a given ethnic community or nation as it exists in a given historical moment. As Todorov put it, "the fear of barbarians is what threatens to make us barbarians. And the evil we will do will be greater than what we feared to suffer. History teaches us: the remedy can be worse than the disease".[39]

Migration policies must be undertaken with the twofold objective of maintaining national and international balances and at the same time of allowing controlled and governed migration flows for those who are forced to migrate. The 2018 Global Compact for Safe, Orderly, and Regular Migration insists on the necessity of adopting international pathways for regular migration.[40] Treaty law can represent a possibility to cope with climate-related international mobility between states with reciprocal and converging interests. Recalling a term coined by the UN

[38] Curtain et al. (2016). Labour Mobility: The Ten Billion Dollar Prize. World Bank. http://documents.worldbank.org/curated/en/171661503669342316/pdf/119105-PUB-PUBLIC-ADD-SERIES-pplabourmobilitybackgroundfinal.pdf. Accessed 14 February 2020.

[39] Todorov. (2009). (op. cit.), p. 16.

[40] UNGA Resolution (2019). Global Compact for Safe, Orderly and Regular Migration (GCM), UN Doc. A/RES/73/195, Objective 5.

Secretary-General's Special Representative on International Migration, the Platform on Disaster Displacement (PDD) talks in this regard about "mini-multilateralism" (or minilateralism), namely, of small group of states with converging interests that can put in place good practices and legal frameworks which might also serve as an example for the entire international community in future.[41] In this sense, one can only rely on the wisdom and foresight of the leaders of the most involved countries to initiate migratory cooperation policies aimed at ensuring a geopolitical balance in the effort to guarantee as much as possible the fundamental rights of the individuals most affected by climate change and who are forced to migrate because of these.

Similar international cooperation agreements actually do already exist[42] and could function as a model and best practice for tomorrow's policies. The Coordination Unit of the Platform on Disaster Displacement highlights, for instance, the role that some bilateral or multilateral migration agreements, like the ones in place under the auspices of Southern African Development Community (SADC), the Economic Community of West African States (ECOWAS), or the European Union (EU).[43] Dedicated agreements might especially be fit for the management of flows coming from micro-state islands. This is the case, for example, of the Pacific Access Category (PAC), established between New Zealand and some Pacific island states. PAC guarantees the right of residence in New Zealand every year to certain quotas of citizens coming from Kiribati, Tuvalu, Fiji, and Tonga. More specifically, it allows the immigration of individuals between 18 and 45 years of age, with job offers in New Zealand and with basic English language proficiency. Interestingly, PAC does not mention climate or environmentally induced migration at any time, probably due to the political difficulties that would have arisen if it did so. On an official level, it represents a classic scheme of interstate migratory cooperation,

[41] Platform on Disaster Displacement (2018). State-Led, Regional, Consultative Processes: Opportunities to Develop Legal Frameworks on Disaster Displacement. In Behrman & Kent (Eds.) (op. cit.), p. 139.

[42] See for instance, New Zealand's five year Strengthened Cooperation Programme with Niue (created in 2004 and reformed in 2009); South Pacific Work Permit Scheme; Pacific Access Category; cooperation agreements adopted under the Pacific Islands Forum.

[43] Platform on Disaster Displacement. (2018). State-Led, regional, Consultative Processes: Opportunities to Develop Legal Frameworks on Disaster Displacement. In Behrman & Kent (Eds.) (op. cit.), p. 139.

thought for migrants who already enjoy some skills and educational qualifications. The "failure" to refer to environmental or climate pressures should, however, not be viewed with concern. Given the uncertainties and the political sensitivity surrounding the issue of "climate migrants", it is probably not necessary for current migration agreements also aimed at preventing forced migrations to focus on a particular migration cause. They can, more easily, guarantee access and residence permits to individuals coming from "environmentally distressed" areas without openly addressing the specific causes of migration. With policies such as PAC, the demographic pressure in island states is progressively diminishing and, at the same time, stable and integrated migration networks, which can be exploited in the future with the worsening of climatic/environmental conditions, are being created. In this sense, debates about the real nature of an agreement like PAC lose relevance: are they special agreements dedicated to "climate refugees" or do they represent a mere intergovernmental cooperation deal on economic migration?[44] If the objective of migration policies aimed at governing migratory flows intensified by climate change is to ensure preventive and voluntary migration from the most affected areas, instruments like PAC, despite their embrionality, can be considered as a promising start.

If, notwithstanding its explicit objectives, PAC may represent a promising start for international cooperation on climate migration, it is also worth noticing that the 2018 Global Compact for Migration explicitly called for the identification and development of pathways and solutions for migrants compelled to leave their countries of origin owing to slow-onset natural disasters, the adverse effects of climate change, and environmental degradation,[45] in an effort to emphasize the environmental or climatic roots of some migration patterns. In a very similar way, the latest Report of the UNFCCC's Task Force on Displacement highlights the government's' opportunity to adapt their national policies on

[44] See in this sense: Williams (2008). Turning the Tide: Recognizing Climate Change Refugees in International Law. *Law & Policy*, 4, p. 515: "It may be argued that the environmental significance of PAC has been exaggerated and that the agreement represents little more than an economically oriented immigration move to bolster New Zealand's workforce, given the variety and class of conditions attached to the category".

[45] Global Compact for Safe, Orderly and Regular Migration, point 21, letter h.

migration so as to "reflect the principles agreed upon at the global level in terms of human mobility in the context of climate change".[46]

Taking a step back, we have seen how the maximalist approach can, if pushed by excessive oversimplification of reality, lean toward policies of closure and cultural containment. In this sense, a minimalist approach, which is more attentive, aware, and informed, should be privileged. This, first and foremost, because it is only in this way that a solid foundation for future legal protection instruments for "climate refugees" can be built. However, the debate between maximalism and minimalism can and should be synthesized, rather than pushed toward a useless polarization and an ensuing intellectual stalemate. A conflicting and misinformed debate between the two schools leads to negative consequences: the two tend to discredit and limit each other rather than to grow together. As Castles put it, "migration scholars must recognize the potential of climate change to bring about fundamental changes in the nature of human mobility, just as environmentalists need to recognize the complex factors that lead some people to adopt migration as a part for their survival strategy".[47]

The minimalist school, whose characteristics have been gradually highlighted in opposition to the maximalist one, underlines the fact that human movements are triggered by complex and multiple causes, among which climate change is just one. It predicts that the number of cases where migration will be directly linked to the effects of climate change will be small. This is corroborated by the fact that most of the affected people will move within their own country, while the maximalist school, as we saw, tends to dramatically highlight the cross-border nature of the phenomenon. Given the weight that economic, ethnic and/or demographic factors can play in the decision of an individual to migrate or

[46] Report of the Task Force on Displacement (TFD Report), Advanced unedited version of 17 September 2018. https://unfccc.int/sites/default/files/resource/2018_TFD_report_17_Sep.pdf. Accessed 14 March 2020, p. 11. It is worth noticing that in the previous versions of the Report, the Task Force more directly ***recommended*** governments to adapt their policies on human mobility in the context of climate change.

[47] Castles (2010). Afterword: What Now? Climate-Induced Displacement After Copenhagen. In McAdam (Ed.) (op. cit.), p. 243.

not, the most radical minimalists even go so far as to conclude that no climatic or environmental danger inevitably leads to migration.[48]

Setting aside this extreme view, minimalism generally believes that the absence of a solid methodological basis can lead to the problems discussed above, which in the end lead to a very low probability of undertaking international migration agreements. Indeed, it is obvious that no interstate agreement on "climate refugees" can be reached if governments rely on "catastrophic" data not enjoying broad scientific consensus. Indeed, while the maximalist literature has advocated the adoption of new international instruments (be it a new treaty, the extension of the 1951 Geneva Convention on refugees, or an extension of the UNFCCC, for example) to provide a dedicated and universal protection regime to "climate refugees", the minimalist literature has encouraged the use of bilateral, minilateral, or multilateral international migration agreements aimed at creating and guaranteeing official migration corridors.

Interestingly, among the limits of minimalism, there is the tendency to neglect the importance of sudden natural disasters in which climate change plays an important role. This category of disasters is at a global level one of the major causes of forced displacement. The IDMC and the Norwegian Refugee Council agree on the fact that natural disasters are the first cause of internal and international displacement worldwide.[49] Climate-related events such as storms, cyclones, floods, hurricanes, etc., *although not caused by climate change, are influenced by it*. Climatology has shown how global warming—albeit with different impacts depending on different geographical areas—is increasing the frequency and strength of extreme meteorological events. Thus, while single events cannot be foreseen, we know they are set to increase in number and intensity due to global warming. It is then risky to conclude that climate change-related migrations cannot exist at all.

In conclusion, the ultimate goal of both maximalism and minimalism is to guarantee additional protection to a category, that of the "climate

[48] Piguet (2008). Climate Change and Forced Migration, New Issues in Refugee Research, Research paper n. 153. UNHCR. https://www.unhcr.org/en-lk/47a316182.pdf. Accessed 13 May 2018, p. 8.

[49] Apap (2019). The Concept of "Climate Refugee": Toward a Possible Definition. Briefing. EPRS | European Parliamentary Research Service. https://www.europarl.europa.eu/RegData/etudes/BRIE/2018/621893/EPRS_BRI(2018)621893_EN.pdf. Accessed 18 December 2019.

refugees", which has come recently to light and that is currently inadequately covered by the international legal system. The two positions are not incompatible[50] and their synthesis is possible. In all this, it is surely improper to use the term "refugees" as an all-encompassing notion covering at the same time migrants, refugees, and asylum seekers. This has contributed over the years to the polarization between the two schools of thought. In reality, they deal with the same topic and aim at the same objective, relying however on radically different investigation methods: maximalism explains with a broad look how and why migration flows—*whatever their nature*—will increase worldwide *also* because of global warming; minimalism pursues this line of research at a different, more in-depth level, trying to better understand the factors, categories, and individuals that are actually at stake. Maximalism and minimalism are not contrasting schools of thought: they investigate two different dimensions of the same phenomenon.

The objective of this book is to understand whether international law provides an obligation to protect individuals compelled to leave their national territory because of anthropogenic climate change. To this end, it follows the minimalist approach, which allows for a more detailed understanding of the (f)actors at play. Within the "climate refugees cauldron", it focuses on a specific minority of "climate refugees", namely, on forced climate migrants.[51] Departing from the maximalist macro level of analysis which fails to address the individual motivations of migration, this book uses the lenses of minimalism to identify which (sub)category of "climate refugees" may actually be considered as legally autonomous under the prism of international protection frameworks. Therefore, the next chapter investigates the panoply of definitions proposed over the years to describe climate-related migrants, with the aim of circumscribing those among them who are actually in need of international protection.

[50] Kälin & Schrepfer (2012). (op. cit.), p. 12.

[51] This book adopts Sumudu Atappattu's definition of forced climate migrants and further builds on it in the following chapters relying on human rights and climate change law considerations. As will be discussed in detail in the next chapter, Atapattu defines forced climate migrants as "people who are forced to leave their homes or lands either temporarily or permanently due to significant environmental damage associated with climate change or where their national state is no longer habitable". See Atapattu (2016). *Human Rights Approaches to Climate Change: Challenges and Opportunities*. New York: Routledge, p. 165.

CHAPTER 5

Defining the Category: Who Are "Climate Refugees"?

As seen, the study of "climate refugees" is rife with different discourses and interpretations that create a certain degree of conceptual and semantic confusion.[1] "It is hard to come up with credible numbers without an agreed definition of climate refugees".[2] Of course, this "definitional chaos" also produces important effects on the legal discussion aimed at identifying viable solutions and applicable regulatory frameworks. For instance, all the scientific and semantic uncertainties that surround our topic have firstly contributed to an exorbitant proliferation of different classifications and proto-legal definitions of the category at hand. Their amount, ranging from the most popular and diffused one, namely, "climate refugees", to more technical ones, such as "climate change-induced migrants", certainly does not help the purposes of a shareable, unitary, and useful legal conceptualization of the phenomenon. Without the identification of a specific category that should be protected under international law in the name of existing rights and obligations, there is the high risk of making climate refugees a hollow concept: it is in fact

[1] For a recent analysis on the different approaches followed in the years to define so-called climate refugees, see Atapattu (2018). A New Category of Refugees? 'Climate Refugees' and a Gaping Hole in International Law. In Behrman & Kent (Eds.), *'Climate Refugees' Beyond the Legal Impasse?* New York: Routledge, pp. 34 ff.

[2] Ibid., p. 36.

G. Sciaccaluga, *International Law and the Protection of "Climate Refugees"*,
https://doi.org/10.1007/978-3-030-52402-9_5

euphemistically incorrect to claim that all environmental, climate, and/or economic migrants enjoy a right to international protection when they leave their countries. The Global Compact for Migration is clear in this regard too, when it states that "*migrants and refugees may face common challenges and similar vulnerabilities*", but that they are indeed "*distinct groups governed by separate legal frameworks*", and that "*only refugees are entitle to the specific international protection defined by international refugee law*" (Preamble, paras. 3 and 4).

Ultimately, international law provides for protection mechanisms only in presence of serious and obvious breaches of fundamental human rights and where the possibility of exercising them in the country of origin of the asylum seeker is precluded. In the absence of a certain threshold of suffering, requests for international protection are normally rejected, with the applicants being merely considered as economic and voluntary migrants, not in absolute need of protection by a third country. By contrast, if some "forced climate migrants" were to enjoy international protection standards, it would be necessary to identify precisely the reasons why these people should be assisted under international law. If such a category, that we may informally call "real climate refugees", exists, its characteristics must thus be clearly explored and defined. In this sense, at least from a legal viewpoint, all those definitions that rely on extremely generic and encompassing approaches are to be discarded. As a way of example, definitions such as "survival refugees", "hunger refugees", or "eco-migrants" (i.e., ecological-economic migrants) are ultimately based on the assumption that, for the purposes of international protection, it does not make much sense to concentrate exclusively on a particular cause of forced migration—climate change in our case—while neglecting others. While ethically and politically admirable, it follows that protection and assistance mechanisms should be created for all those who migrate internationally in situations of existential distress, irrespective of the actual cause that determined the choice to migrate. While appreciating the ethical value of this line of reasoning, this book takes distance from it both because it is unlikely to be politically realistic and because it is based on questionable legal grounds.

As said, a shared legal conceptualization of whom we commonly refer to as climate refugees is still lacking. This leads to several problems in the path toward the identification of potentially applicable regulatory frameworks. Hence, there is a strong need to clarify the terminology,[3] with the precise intention of understanding which "climate refugees" among the many that are in our "cauldron" could effectively enjoy a right of protection under international law. To do this, it is first useful to investigate the different definitions and classifications that have been proposed over the years by many authors, including both academics and practitioners.

Chapter 3 mentioned the typology adopted by Renaud,[4] which is very valuable for the purposes of a correct conceptualization and understanding of the phenomenon. In addition to Renaud's, 2012 Kälin and Schrepfer's classification is useful as well. It identifies different sorts of climate migrations differentiating them according to the diverse environmental/climatic event at their origin, also with the objective of better understanding which migrations require which answers, both at the political and at the legal level.[5]

[3] In this sense, see Nansen Conference on Climate Change and Displacement in the 21st Century (2011). Chairperson's Summary. https://www.unhcr.org/4ea969729.pdf. Accessed 12 January 2020, para. 21.

[4] Please refer to Chapter 3

[5] Kälin & Schrepfer (2012). Protecting People Crossing Borders in the Context of Climate Change: Normative Gaps and Possible Approaches. UNHCR Legal and Protection Research Series, PPLA/2012/01. https://www.unhcr.org/4f33f1729.pdf. Accessed 12 December 2016. We here report only part of Kälin and Schrepfer's categorization of environment-related migrations, by focusing only on those parts that are relevant to the present work. For seek on completeness, Kälin and Schrepfer's categorization also highlights the cases in which a government declares a certain area to be at high risk and too dangerous to sustain human life due to environmental dangers. Consequently, the inhabitants of the area would be forced to evacuate and relocated to other places, with the prohibition of returning to their homes of origin. This type of case will affect a relatively small number of people, but could raise significant legal complexities; it is also strongly connected to the c.d. relocations induced by the construction of large infrastructure projects, such as dams, highways, power plants, etc. In such hypotheses, it is necessary to underline how the displacement of the persons involved would be, if only in theoretical line, exclusively internal; it remains, therefore, to understand how and how much international law can, if not through the adoption of guidelines and exhortative tools, be incisive in similar cases. Finally, the events likely to undermine the internal public order of a State, the cases of endemic violence and even of armed conflict that could be caused, at least indirectly, by an impoverishment of essential resources caused by climate change remain to be considered.

i. Kälin and Schrepfer first draw attention to *sudden-onset disasters* (floods, windstorms, landslides, heavy rain, and so on), events capable of triggering forced displacements and generating high economic costs. Depending on the vulnerability and resilience of the affected communities, and on the efforts aimed at the recovery of the affected areas, the displacement of populations in reaction to these disasters tends to be temporary, since return remains in most cases possible.
ii. Secondly, they consider gradual and slow environmental degradation, so-called *slow-onset disasters*. Among these, we may think of sea-level rise (with the consequent "deterritorialization" and the salinization of the water supplies), desertification, soil erosion, the long-term effects of recurrent floods, the dissolution of permafrost,[6] heat waves, and droughts. Similar environmental changes have negative effects on economic opportunities and living conditions in the affected areas. Although such effects do not necessarily provoke forced migrations, they can certainly induce people to consider migration as a way to adapt to environmental changes. That said, should the affected country become uninhabitable over time due to very serious degradation, the affected populations would be forced to migrate permanently. It is here of course that the infamous "sinking" small island states become very important. In fact, as a consequence of rising sea levels and, therefore, due to widespread soil erosion, increasingly invasive tides (with the resulting salinization of aquifers), rising water temperatures (with the resulting alteration of island and marine ecosystems and damage to coral reefs), their territories could become uninhabitable over the next decades or, in extreme cases, even disappear altogether. At that point, the concerned populations would be forced to move abroad permanently. This creates problems not only for the protection and management of the individuals involved, but also for their enjoyment of collective rights, such as their self-determination as peoples in the event of the disappearance of a state's territory. These problems would be caused (mainly) by the activities of other states that have been responsible for greenhouse gas emissions while being

[6] See, for instance, in the Italian newspaper *La Repubblica*: https://www.repubblica.it/ambiente/2019/09/03/news/esperto_artico_russo_la_condizione_del_permafrost_e_critica-235071230/. Accessed 23 March 2020.

aware of their effects. Similar geographical contexts, although representing isolated systems, are interesting windows on the future large-scale impacts of climate change and their legal implications.[7]

Further on, an alternative typology of "climate refugees" is the one proposed by Barnett and Webber. To understand the phenomenon from a legally useful point of view, this typology has the merit of highlighting that most climate refugees are in fact "simple" labor migrants, part of so-called invisible migrations, not likely to activate incisive international protection and assistance standards.[8]

i. The first subset of the typology consists of "*migrant workers, international and not*": individuals exposed to the risk of the negative effects of climate change that deliberately decide to migrate to adapt to the new context. Given that migrant workers are generally young people, able to rely on a welcoming social network in the destination country, it is reasonable to assume that this group of people does not, for the most part, require significant international protection in a foreign country. It would remain however appropriate to create and implement as much as possible international migration policies and corridors aimed at facilitating climate-induced migration.
ii. Secondly, Barnett and Webber talk of "*forced displacements, international and not*". As known, the severity and frequency of sudden meteorological disasters is increasing due to global warming. Migrations associated with similar events are thus becoming more and more conspicuous and frequent. It is necessary for governments and the international community to increase their capacity of planning and managing similar situations (thereby investing on disaster

[7] For further information on these contexts, see *ex multis*, the following study in environmental matters: Albert et al. (2016). Interactions Between Sea-Level Rise and Wave Exposure on Reef Island Dynamics in the Solomon Islands. *Environmental Research Letters*, 11. We report here the following and significant words: "We have documented five vegetated reef islands (1–5 ha in size) that have recently vanished [...]. Shoreline recession at two sites has destroyed villages that have existed since at least 1935, leading to community relocations".

[8] Barnett & Webber (2010). Migration as Adaptation. In McAdam (Ed.), *Climate Change and Displacement, Multidisciplinary Perspectives*. Oxford: Hart Publishing, pp. 42–43.

risk reduction), as well as to enhance their ability to provide humanitarian assistance where needed.

iii. Then, "*permanent migrants, international and not*" come into play: due to gradual environmental degradation caused by climate change, more and more people are and will be pushed to migrate permanently. These persons need, at least theoretically, more assistance, especially if their home states become unable to guarantee the exercise of fundamental rights in their territories because of global warming.

This last typology is essentially structured upon two dichotomies: the compulsory/voluntary nature and the temporary/permanent one of the migrations in question. The forcedness and duration of human migration are indeed decisive elements also when the discussion concerns asylum and international protection law. So, to further clarify the picture, it is important to distinguish between climate-related "events" and "processes". The former consist of sudden events that, in a more or less significant way, are characterized by a short duration and by considerable intensity. On the contrary, climate-related "processes" are gradual and progressive. In an international law analysis, this distinction is of course not to be underestimated: different phenomena require different answers. Migrations induced or caused by climate processes present in fact greater complexities and challenges. For sudden-onset disasters there is already an applicable regulatory framework (even if it is to be developed and further implemented) within the (inter)national laws and practices envisaged for the management of natural disasters. By contrast, the field of gradual environmental degradations associated with climate change concerns a domain that is not currently addressed at a legal, political, or practical level.

Both Kälin's and Barnett's classifications help to better understand a phenomenon that, despite the simplistic (ab)use of the term "climate refugees", is actually very complex. They furthermore provide guidance toward the critical investigation of the panoply of (proto)legal definitions of climate-related migrants that crowd today the studies and the discussions on the issue.

As well known, the most commonly used term in our domain is "climate refugees". It originates from a broader group—that of "environmental refugees"—which in turn is a term probably used for the first time in 1976 by the founder of the Worldwatch Institute, Lester Brown.[9] This terminology was taken up again in the early Eighties by those who began exploring the links between development, conflicts, environmental degradation, and human migrations.[10] The term "environmental refugee" then emerged in the academic and international political debate after 1985, above all thanks to the work of Essam El-Hinnawi, whose terminological approach has been followed by the majority of authors in the following years.[11] According to El-Hinnawi's definition, adopted and published by UNEP in 1985, environmental refugees are "are those people who have been forced to leave their traditional habitat, temporarily or permanently, because of a marked environmental disruption (natural and/or triggered by people) that jeopardized their existence and/or seriously affected the quality of their life. By 'environmental disruptions' in this definition is meant any physical, chemical and/or biological changes in the ecosystem (or the resource base) that render it, temporarily or permanently, unsuitable to support human life".[12]

[9] Truly speaking, a reference was made to the notion of "ecological refugees" already in 1948. See on this: Vogt (2011). Road to Survival. Cit. in Gemenne, How They Became the Human Face of Climate Change: Research and Policy Interactions in the Birth of the "Environmental Migration" Concept. In Piguet, Pécoud, de Guchtneire (Eds.), *Migration and Climate Change* (p. 227). Cambridge: Cambridge University Press and UNESCO Publishing.

[10] See as a matter of example, Hillary (Ed.) (2000). *Ecology 2000: The Changing Face of Earth, Joseph.* New York: Beaufort Books, p. 214. See also Timberlake (1983). *Environmental Wars and Environmental Refugees: The Political Background to the Cartagena Convention.* London: Earthscan. For an in-depth analysis on the evolution of the terminology used regarding the category of "climate refugees", see Terminski (2012). Toward Recognition and Protection of Forced Environmental Migrants in the Public International Law. Refugee or IDPs Umbrella? https://papers.ssrn.com/sol3/papers.cfm?abstract_id=2029796. Accessed 14 September 2019.

[11] Cf Westing (1992). Environmental Refugees: A Growing Category of Displaced Persons. *Environmental Conservation*, 19, pp. 201 ff. See also Myers (1993). Environmental Refugees in a Globally Warmed World. *Bioscience*, 43, pp. 752 ff. Shurke & Visentin (1992). The Environmental Refugee: A New Approach, Ecodecision. *Canada's Periodical on Refugees*, 1, p. 1 ff. Keane (2004). The Environmental Causes and Consequences of Migration: A Search for the Meaning of 'Environmental Refugees'. *Geo. Int'l Law Review*, 2, pp. 210 ff.

[12] El-Hinnawi (1985). Environmental Refugees. Report. UNEP. https://digitallibrary.un.org/record/121267. Accessed 12 February 2020. It is worth recalling that El-Hinnawi was not trying to raise any kind of legal, or even ethical, argument about extending refugee law to persons displaced for environmental reasons. Rather, he merely used the term "refugee" to highlight

Over the years, this definition has had a significant impact on the political and academic communities, with variants of this same labelling being commonly used. On the furrow of what was done by in El-Hinnawi's, these attempts, remaining at a very general level, have probably more merits from a descriptive than from a legal point of view. Norman Myers for example defined environmental refugees as "people who can no longer gain a secure livelihood in their homelands because of drought, soil erosion, desertification, deforestation and other environmental problems, together with the associated problems of population pressures and profound poverty".[13] Other authors have not deviated markedly from this level of generality. Ben Gorlick, former Senior Policy Advisor of the United Nations, defined environmental refugees as "people who are displaced from or who feel obliged to leave their usual place of residence, because their lives, livelihoods, and welfare have been placed at serious risk as a result of adverse environmental, ecological, or climatic processes and events".[14]

Definitions of this kind are useful to describe in general terms the phenomenon at hand and the category of people displaced by environmental degradations, but they are excessively vague if one wants to extricate oneself from the climate refugees cauldron. We saw that it is very difficult to understand precisely when human migrations are actually caused by environmental degradations solely, not to talk about climate change, and this difficulty alone already undermines the legal utility of such definitions. The complete lack of consideration given to the distinction between voluntary and forced movements, as well as to the distinction between internal and international ones, makes the notion of "environmental/climate refugee" far too vague and generic. Following this reasoning, some scholars have even questioned the usefulness *tout*

the potentially devastating impact of uncontrolled development and pollution. file:///C:/Users/ius3/Downloads/Environment_and_forced_migration_%20(1).pdf. Accessed 17 February 2020.

[13] In Myers (2001). Environmental Refugees: A Growing Phenomenon of the 21st Century. *Philosophical Transactions of the Royal Society: Biological Sciences*, 357, p. 609.

[14] van Wormer &. Beshorn (2011). *Human Behavior and the Social Environment. Groups, Communities and Organizations*. Oxford: Oxford University Press, p. 279.

court of the notion itself, condemning its excessive generality.[15] Unsurprisingly, this terminological approach has also been criticized by the United Nations High Commissioner for Refugees (UNHCR), that has expressed several times serious reservations about the use and dissemination of the term "environmental refugee", since it does not enjoy any link with international refugee law and because it could even potentially undermine such regime by making states increasingly skeptical of it and thus less willing to apply it.[16] Quite curiously then, this approach is even opposed by those countries that are most vulnerable to global warming, namely, by those micro-states that at risk of "drowning" and becoming deterritorialized. In Kiribati and Tuvalu, for example, the term "climate refugee" is openly rejected both by the government and by the population, because it implies a certain degree of impotence and lack of dignity (as well as dependence on foreign aid), with which the populations do not intend to see themselves depicted.[17]

Due to these flaws, the term "environmental/climate refugee" must be considered with due care. The implication behind the word "refugee", when put in relation with environmental or climate factors is that the kind of migration considered is produced by such factors in a *vacuum*, and that it is forced and international. In brief, the underlying reasoning

[15] Bates (2002). Environmental Refugees? Classifying Human Migrations Caused by Environmental Change. *Population and Environment*, 5, p. 466.

[16] See in this sense Ogata (former United Nations Commissioner for Refugees) declaration released in Geneva in October 1992 during a meeting of the Swiss Peace Foundation: "Using the term 'environmental refugee' to refer to all people forced to leave their homes because of environmental change loses the distinctive need of refugees for protection. It blurs the respective responsibilities of national governments towards their citizens and of the international community towards those who are without protection. It also impedes a meaningful consideration of solutions and action on behalf of the different groups. Therefore, UNHCR believes the term 'environmental refugee' is a misnomer". https://www.unhcr.org/admin/hcspeeches/3ae68fcc1c/statement-mrs-sadako-ogata-united-nations-high-commissioner-refugees-luncheon.html. Accessed 25 March 2019.

[17] The former Kiribati President words on this better explains than any other the refusal to rely on the label "climate refugee": "When you talk about refugees – climate refugees – you are putting the stigma on the victims, not the offenders [...]. We do not want to lose our dignity. We are sacrificing much by being displaced, in any case. So we do not want to lose that, whatever dignity is left. So the last thing we want to be called is 'refugee'. We are going to be given as a matter of right something that we deserve, because they have taken away what we have", cit. McAdam (2012). *Climate Change, Forced Migration, and International Law*. Oxford: Oxford University Press, p. 41.

of the word refugee is not supported by any legal basis in these cases,[18] consequently generating semantic chaos due to its excessive generality. On top, it is even rejected by the populations that should, theoretically, benefit from it. However, notwithstanding all these shortcomings, the term "environmental/climate refugee" remains extremely effective from a communicative and symbolic point of view, and this is simply because it immediately tells a story about the state of necessity and the underlying causes of an emerging and increasingly relevant category of migrants.[19]

In order to remedy these flaws, a further, more technical and less evocative definition has been proposed over time: that of "environmentally displaced person". Today, while the wording "climate refugee" is often used in the media given its communicative and symbolic power, the term "environmentally displaced person" tends to be favored by the various international organizations called upon to deal with the phenomenon. This terminology has in fact been adopted by the UNHCR, the Office of the High Commissioner for Human Rights (OHCHR), the IOM, and the Refugee Policy Group, as well as by the UN Secretary-General. According to IOM, for example, environmentally displaced persons are those "who are displaced within their country of habitual residence or who have crossed an international border and for whom environmental degradation, deterioration or destruction is a major cause of their displacement, although not necessarily the sole one".[20] In an extremely general way again, an environmentally displaced person has also been defined as "one who leaves his or her home and seeks refuge

[18] As Mayer explains: "By implying that migration was forced and international, the term "refugee" was analytically misleading". In Mayer (2018). Who Are "Climate Refugees"? Academic Engagement in the Post-truth Era. In Behrman & Kent (Eds.), *'Climate Refugees' Beyond the Legal Impasse?* New York: Routledge, p. 93.

[19] Quite surprisingly (and radically), for instance, Gemenne has recently pointed out that the category at stake, notwithstanding the legal incorrectness of using the term refugee, should still be labeled as that of "climate refugees", since this choice recognizes in the best way possible "that these migrations are first and foremost the result of a persecution that we are inflicting to the most vulnerable". See Gemenne (2017). The Refugees of the Anthropocene. In Mayer & Crépeau (Eds.), *Research Handbook on Climate Change, Migration and the Law*. Cheltenham: Elgar, p. 394.

[20] IOM (2019). Environmentally Induced Population Displacements and Environmental Impacts Resulting from Mass Migration (p. 4). International Symposium, IOM. https://publications.iom.int/system/files/pdf/environmentally_induced.pdf. Accessed 12 December 2019.

elsewhere for reasons relating to the environment".[21] A more precise definition is the one proposed in the Draft Convention on the International Status of Environmentally Displaced Persons, drafted by scholars of the University of Limoges. It talks of "individuals, families and populations confronted with a sudden or gradual environmental disaster that inexorably impacts their living conditions and results in their forced displacement, at the outset or throughout, from their habitual residence and requires their relocation and resettlement".[22]

In general, this new definitional approach was intended, first of all, to distinguish between refugees (in the legal sense of the term) and persons induced or forced to leave their habitual place of residence due to environmental degradation.[23] In this sense, for example, Hens argued in his studies on environmental-induced migration in Ghana that "persons who are displaced within their country of habitual residence or who have crossed an international border and for whom environmental degradation, deterioration or destruction is a major cause of their displacement, although not necessarily the sole one, belong to the environmentally displaced people. These persons are refugees in the real sense of the word, but their situation does not coincide with the legal definition of refugee".[24]

In essence, there is one common basic element linking both the term "environmental refugee" and "environmentally displaced person", and

[21] Falstron (2001). Stemming the Flow of Environmental Displacement: Creating a Convention to Protect Persons and Preserve the Environment. *Colorado Journal of Environmental Law and Policy*, 15, p. 1 ff.

[22] Prieur et al. (2013). Draft Convention on the International Status of Environmentally Displaced Persons (third version). University of Limoges. https://cidce.org/wp-content/uploads/2016/08/Draft-Convention-on-the-International-Status-on-environmentally-displaced-persons-third-version.pdf. Accessed 12 January 2020.

[23] A similar all-inclusive approach, albeit limited to internal displacement, is also followed by the Peninsula Principles on Climate Displacement, that define "climate displacement as movement of people within a State due to the effects of climate change, including sudden and slow-onset environmental events and processes, occurring either alone or in combination with other factors". See Principle 2 of the Peninsula Principles on Climate Displacement within States, Displacement Solutions, 18 August 2013. http://displacementsolutions.org/peninsula-principles/. Accessed 12 February 2020.

[24] Hens (2012). Environmentally Displaced People. *Encyclopedia of Life Support Systems, Area Studies Africa (Regional Sustainable Development Review)—Vol. II*. http://web.mnstate.edu/robertsb/308/environmentally%20displaced%20people.pdf. Accessed 14 May 2019.

this is the compulsory nature of the migration. These definitions therefore neglect the group of those who are "only" induced or encouraged to leave their home due to environmental or climate pressures. As seen, although it is often very difficult to draw a clear line of distinction between voluntary and forced migrants, the legal implications of this distinction are indeed decisive. This implies that it is impossible to compare a forced migrant to a voluntary one: while the former—not enjoying the chance of remaining in his or her origin country—should count on the assistance and protection of a third state or the international community as a whole, the voluntary migrant, still able to rely on the potential and effective help of his or her national state, is not in a state of extreme necessity and therefore cannot claim a right to be protected in a third state. For this reason, the definitions of the category at hand that do not adequately take into account the distinction between obligatory and voluntary movement and, at the same time, between internal and international mobility, are not useful for the purposes of a study trying to identify useful criteria for the creation or evolution of international legal instruments aimed at guaranteeing international protection to forced migrants.

The notions of "environmental refugee" or "environmentally displaced person" are based, ultimately, on the idea of a forced and compulsory displacement. Other approaches followed over the years have instead leaned toward identifying those who decide on more or less voluntary bases to abandon their homes also due to environmental or climate pressures. When this happens, the term "migrant" is used more often, entailing a certain degree of voluntariness that is normally not at the disposal of a refugee, understood in the broadest sense. In 2008, the IOM itself revised its position, and provided an updated definition that also includes those who migrate in a "voluntary" way. The term "environmental migrants" was therefore proposed, so as not to focus exclusively on forced movements. Environmental migrants are "persons or groups of persons who, for compelling reasons of sudden or progressive changes in the environment that adversely affect their lives or living conditions, are obliged to leave their habitual homes, or choose to do so, either temporarily or permanently, and who move either within their country or abroad".[25]

[25] Opening Statement by McKinley, IOM General Director, in occasion of the *Conference on Climate Change, Environmental Degradation and Migration: Addressing Vulnerabilities and Harnessing Opportunities*. Geneva, 19 February 2008, p. 43.

A similar definition certainly has considerable descriptive merits, but it should also carefully looked at, since it would be very complex and probably unfeasible to create all-encompassing protection policies and legal instruments, capable of providing assistance both to movements that are still partly voluntary and to those that are totally forced, as well as to temporary and permanent, internal and international ones. This definition has been strongly criticized by Kälin and Zetter, for instance, who affirm that it could undermine the legal framework built for Internally Displaced Persons (IDPs),[26] given the absence of a distinction between international and internal migrations. Furthermore, neglecting the difference between voluntary and forced displacement, which is crucial from a regulatory viewpoint, this definition does not allow us to understand which categories should be protected and which not. Thus, in summary, "while it is true that social scientists often use migration as a generic term encompassing both voluntary and forced movements, international law does not use the term 'migrant' in the context of forced movements but refers to 'displaced persons' and 'refugees'".[27]

If this is true, however, it is probably also incorrect to adopt a completely dichotomous approach between the "sociological" and "legal" notions of refugee. And Mayer beautifully expresses the reason why: "if one follows Bourdieu's perspective on the use of language as a symbolic power, the [legal] misnomer argument appears particularly conservative - a reaffirmation of a given power relation in favour of the dominant classes. On the other hand, labelling environmental migrants 'refugees' may contribute to emancipatory strategies for their protection. If social notions may evolve, so is also the case of legal ones. There is a *continuum*

https://publications.iom.int/system/files/pdf/hsn_quadrilingual_report.pdf?language=en. Accessed. 15 February 2020.

[26]The term Internally Displaced Person, as defined in the 'Guiding Principles on Internal Displacement' (Principle No. 2), refers to "persons or groups of persons who have been forced or obliged to flee or to leave their homes or places of habitual residence, in particular as a result of or in order to avoid the effects of armed conflict, situations of generalized violence, violations of human rights or natural or human-made disasters, and who have not crossed an internationally recognized State border". See UN High Commissioner for Refugees (UNHCR) (1998). Guiding Principles on Internal Displacement. ADM 1.1.PRL. 12.1, PR00/98/109.

[27]See Kälin & Schrepfer (2012) (op. cit.), p. 29. See also Zetter (2010). Protecting People Displaced by Climate Change: Some Conceptual Challenges. In McAdam (Ed.) (op. cit.), pp. 140 ff.

rather than a dichotomy between the 'legal' and the 'sociological notion of refugee. [...]. The 'legal misnomer' argument is based on a wrong understanding of law as an immutable set of given norms. [...]. Legal notions are and need to be open to renegotiation".[28]

Indeed, refugee law has to a certain extent surely evolved since the adoption of the Geneva Convention on the Status of Refugees in 1951. As known, within the African and Latin American spheres, for instance, the definition of refugee has been expanded to include those fleeing because of war, occupation, or events capable of seriously disturbing the public order.[29] This means that, although we always have to take into account the centrality of the refugee definition of the 1951 Geneva Convention, the category of (legal) refugees is not immutable at all under international law. In fact, it adapts and may further adapt in future in light of social and historical phenomena that need new legal responses. Nonetheless, such evolutions occur slowly, since the international community needs time to reach new compromises in a complex system of values, priorities, and interests.

Turning back to the panoply of existing definitions, the one of "forced climate migrants" proposed by Atapattu in 2016 is of particular interest. In addition to focusing exclusively on forced, cross-border migrations, it also briefly indicates a course of action that the receiving states should undertake to guarantee the enjoyment of basic rights of the individuals

[28] Mayer (2012). "Environmental Refugees"?: A Critical Perspective on the Normative Discourse. Center for Sustainable Development Law. https://papers.ssrn.com/sol3/papers.cfm?abstract_id=2111825. Accessed 14 January 2016, p. 4.

[29] For the Americas, see the 1984 Cartagena Declaration on Refugees, adopted by the Colloquium on the International Protection of Refugees in Central America, Mexico and Panama, held at Cartagena, Colombia from 19–22 November 1984, point 3: "Hence the definition or concept of a refugee to be recommended for use in the region is one which, in addition to containing the elements of the 1951 Convention and the 1967 Protocol, includes among refugees persons who have fled their country because their lives, safety or freedom have been threatened by generalized violence, foreign aggression, internal conflicts, massive violation of human rights or other circumstances which have seriously disturbed public order". For the African continent, see the Convention Governing the Specific Aspects of Refugee Problems in Africa ("OAU Convention"), adopted on September the 10th 1969, entered into force on June the 20th 1974, article 1.2: "The term 'refugee' shall also apply to every person who, owing to external aggression, occupation, foreign domination or events seriously disturbing public order in either part or the whole of his country of origin or nationality, is compelled to leave his place of habitual residence in order to seek refuge in another place outside his country of origin or nationality".

involved. "Forced climate migrants" are "people forced to leave their homes or territories, temporarily or permanently, due to significant environmental damage linked to climate change or due to the uninhabitability of their state of origin. The receiving state should guarantee respect for the basic human rights of forced migrants until the return to their homeland becomes possible again (in the case of temporary displacement), while appropriate agreements are to be made to ensure integration and residence in the receiving country [in the case of permanent migrants]".[30]

In conclusion, it is clear that a first and important gap in our field is the absence of a shared terminology (not to mention, therefore, of a legally binding one) concerning which "climate refugees" should be protected. The Nansen Initiative has underlined this deficiency and has reiterated that the term "environmental/climate refugee" is inaccurate and misleading.[31]

More recently, the terminological debate concerning the phenomenon of climate/environmental forced migration has begun to focus more on the notion of "disaster". In this regard, for example, the Nansen Initiative published in 2015 the policy document named "Agenda for the Protection of Cross-Border Displaced Persons in the Context of Disasters and Climate Change" (the Protection Agenda). The Agenda is publicly supported by 109 states, its main purpose is to improve the understanding of the phenomenon, as well as to identify the most effective practices, both internationally and nationally, that exists today with regard to the assistance of those who are forced to leave their country due to natural disasters.[32] Of course, the notion and definition of disaster

[30] Atapattu (2016). *Human Rights Approaches to Climate Change: Challenges and Opportunities*. New York: Routledge, p. 165.

[31] An important intergovernmental initiative sponsored by the Swiss and Norwegian governments, which was attended by over one hundred States with the intention of discussing the issue of climate and environment-related migration and to provide political guidelines to follow. For further information, refer to the website www.nanseninitiative.org.

[32] As per protection, the Agenda refers to "any positive action, whether or not based on legal obligations, undertaken by States on behalf of disaster-displaced persons or persons at risk of being displaced that aim at obtaining full respect of the rights of the individual in accordance with the letter and spirit of the applicable bodies of law, namely, human rights law, international humanitarian law and refugee law". See The Nansen Initiative (2015). Agenda for the Protection of Cross-Border Displaced Persons in Context of Disasters and Climate Change Volume I, para. 14. https://nanseninitiative.org/wp-content/uploads/2015/02/PROTECTION-AGENDA-VOLUME-1.pdf. Accessed 19 March 2020.

becomes thence crucial. The Nansen initiative has followed the definition proposed by the UN International Strategy for Risk Reduction (UNISDR), according to which a disaster is a "serious disruption of the functioning of a community or a society involving widespread human, material, economic or environmental losses and impacts, which exceeds the ability of the affected community or society to cope using its own resources".[33] According to the Protection Agenda, disaster displacement then refers to "situations where people are forced or obliged to leave their homes or places of habitual residence as a result of a disaster or in order to avoid the impact of an immediate and foreseeable natural hazard".[34]

If "disaster" is a crucial notion for the Nansen Initiative and its follow-up, the same is not for the Paris Agreement on Climate Change, adopted in 2015 as well. Where the Agreement refers to climate-related migrations, the reference to the term disaster is missing. In fact, the Paris Agreement is quite soft on the subject matter: only in the Decision—the preliminary and non-binding part of the agreement—we may find an explicit reference to what is labeled as "displacement related to the adverse impacts of climate change". The Decision provides that the Warsaw International Mechanism (WIM), the system under the UNFCCC auspices that deals with the study and development of the "loss and damage" section,[35] shall create a team of experts and representatives of the most relevant organizations in the field, with the objective of addressing and minimizing the risks associated with the phenomenon.[36] The work of

[33] UN Office for Disaster Risk Reduction (2009). UNISDR Terminology on Disaster Risk Reduction. https://www.unisdr.org/files/7817_UNISDRTerminologyEnglish.pdf. Accessed 15 May 2019.

[34] Agenda for the Protection of Cross-Border Displaced Persons in Context of Disasters and Climate Change Volume I, para. 16.

[35] Interestingly, it has been possible to insert a section on loss and damage in the Paris Agreement only because it is accompanied by a provision in the Agreement's Decision establishing that the interested article "*does not involve or provide a basis for any liability or compensation*". See para. 51 of Decision 1/CP.21 (2015), Adoption of the Paris Agreement.

[36] Decision 1/CP.21, Adoption of the Paris Agreement (UN Doc. FCCC/CP/2015/10/Add.1), 2016, para. 39: "Also requests the Executive Committee of the Warsaw International Mechanism to establish, according to its procedures and mandate, a task force to complement, draw upon the work of and involve, as appropriate, existing bodies and expert groups under the Convention including the Adaptation Committee and the Least Developed Countries Expert Group, as well as relevant organizations and expert bodies outside the Convention, to develop recommendations

this Task Force will be analyzed later in this book.[37] It suffices here to highlight that the different definitions followed by the most recent instruments adopted by the international community still show the need for further and significant studies, aimed first of all at providing a conceptualization of the category of "climate refugees" that may turn out to be shared and thus useful for the purposes of international law.

Concisely, we may conclude that the choice of the terminology used, ascribing to it different meanings and implications, appears ultimately quite arbitrary. For example, if the word migrant is often used to imply some voluntariness in the migratory event, while the word refugee implies an obligatory nature, it is also true that often the terms are also used in the opposite way. Renaud's classification for example remains one of the most accurate and useful from a conceptual point of view even if it is based exclusively on the word "migrant". Thanks to the use of appropriate specifications, Renaud provides an exhaustive classification of environment-related human migrations while not actually distancing himself from the word migrant. Thus, it is the clarity of the different authors that determines which terminology makes more or less sense.

To offer a reasoning that might prove useful to international law studies, this book looks within the "cauldron of climate refugees" and tries to distinguish which persons, among this confused group, may be actually entitled to international protection rights. We saw that a first basic distinction capable of shedding some light is the forced/voluntary displacement dichotomy. When taken into account, this distinction makes it possible to distinguish between those who are in a situation of real need for international protection and those who, on the contrary, must be included in normal, economic migration policies. This is a crucial point, and should not be underestimated. Maintaining, in the name of utopian humanitarian principles, that every international migrant leaving his or her country because of difficulties experienced therein enjoys international protection rights in other countries is just wrong.[38] Of course, it is often

for integrated approaches to avert, minimize and address displacement related to the adverse impacts of climate change".

[37] Please refer to Chapter 10.

[38] In this regard, Mayer interestingly argues the following: "[…] what international, regional or domestic laws and institutions call 'refugee' is only the tip of the iceberg of migrants in dire need of international protection – and yet, even among those States which did ratify the 1951 Convention relating to the Status of Refugees and its protocol,

difficult to draw a line on the *continuum* we talked of about voluntary and forced migrants, but it remains nonetheless crucial in a legal effectiveness reasoning. "Traditional" refugees must provide the authorities of the receiving state with sufficient evidence as to the existence of a risk of persecution in their country of origin. If this risk is found to be true, the asylum seeker is a forced migrant, *non refoulement* obligations apply and he or she must be protected by the receiving state. A similar mechanism should also apply when individuals migrate because of climate change: if they can prove that life in their country of origin has become effectively intolerable, then they should be protected as forced climate migrants (let us say as "real climate refugees"), but if they cannot, it is not a matter of international protection anymore.

In what follows, this book will focus on the "protection-side" of the phenomenon rather than on the "migration management" one. We have already discussed how climate change and consequent environmental degradations can essentially generate two types of displacement: a forced one, as a consequence of the impossibility for the individual to remain in the own state of origin due to its environmental conditions; the other voluntary, resulting from environmental and climate processes that, although potentially serious in the long run, lack, in the short run, of a force and incisiveness capable of immediately activating international protection mechanisms in a third country. For this second group of migrants it is necessary, where possible, to create and implement far-sighted migration policies, so as to ensure that migration flows from the regions most affected by phenomena such as desertification and rising ocean levels are managed in respect of the dignity of the persons and communities involved, both emigrating and receiving ones. For this group, however, it is risky to advocate a solid right to international protection. In fact, current and relevant international protection instruments, whose cornerstone remains the 1951 Refugee Convention, tend

many, most perhaps, are in patent breach of their obligation. The international climate regime appears more and more to conform to its depiction by James Hathaway in the 1990s as an attempt by sovereign States 'to govern disruptions of regulated international migration in accordance with the interests of States', rather than as a humanitarian project of sorts. The recognition of a relatively tiny population of refugees has made it more legitimate for States to repress many other force migrants on arbitrary bases". Mayer (2017). Climate Change, Migration and the Law of State Responsibility. In Mayer & Crépeau (Eds.), *Research Handbook on Climate Change, Migration and the Law*. Cheltenham: Elgar, p. 239.

to exclude that international protection can be guaranteed where certain requisites are missing. Among these, an important role, in addition to that pertaining to serious violations of primary rights, is played by the imminence of the harm feared. To date, if the danger perceived by the person requesting international protection is not imminent, dedicated regulatory systems do not guarantee the recognition of the necessary statuses.

Further on, the book will focus therefore on the protection to be provided to those that we will define from now on as "forced climate migrants" (following Atapattu's definitional approach), namely, individuals forced to leave their country due to serious environmental degradation generated by anthropogenic climate change, and we are going to focus on the legal arguments on the basis of which this category should be specifically treated and protected under international law.

Before proceeding in this direction, however, a final and very important conceptual consideration is of order. If, as pointed out, a "forced climate migrant" becomes such when forced to leave his or her state of origin, it is necessary to dwell carefully on what literature calls "reasonable internal flight alternative".[39] Generally speaking, it is possible to argue that an individual should be protected by the international community and a third state in the event that he or she does not enjoy the possibility to avail him/herself of the assistance and support of the home state, as it is exactly the case under refugee law and the 1951 Geneva Convention. In this case, the international legal order, in the name of the protection of universally recognized human rights, acts and provides for some sort of protection. However, when an individual still enjoys a living and existing bond of trust and cooperation with his/her country, the reasons why a third state should be obliged to protect that very same individual become unclear. In this sense, the importance of internal flight alternatives should not be underestimated. Generally, a refugee (from a legal perspective of course) escapes from his or her own state due to a particular conduct (or omission) of the latter and, thus, flees from the political landscape of the country of origin. If a climate-related migrant is forced to leave a specific area because this has become uninhabitable, this does not mean that he or she is in the automatic need to leave the country of origin completely. When an internal flight is available, international protection

[39] Mayer (2012) (op. cit.), pp. 5–6.

can be easily excluded.[40] Logically, therefore, we should talk of forced climate migrants only when the internal migration option is reasonably precluded. By so doing, our "cauldron" starts becoming far smaller and more easily discernible, thanks to more precise definitional boundaries.

Some decades ago, Anthony Shacknove identified the limits of the 1951 Convention definition of refugee just by wondering what a refugee was. Interestingly, in his opinion a refugee was then every person for whom the social contract between the state and citizen has broken down. This is the reason why this person must seek refuge in another state.[41]

So, as there is a clear legal distinction between a refugee and an IDP for example, the very same distinction should stand between internal and international forced climate migrants. It should be noted, in the end, that the eventualities in which a reasonable internal flight alternative is or will be effectively precluded because of climate change are in fact very limited. This might occur above all in those insular micro-states whose territory is wholly—although in the medium/long and not short-run—destined to become uninhabitable. It would also apply to those cases where climate or environmental pressures impact on already extremely difficult demographic and socio-economic contexts, generating individual situations in which the human dignity of a hypothetical asylum seeker in a third state would be infringed in the event of a return (subsequent to *refoulement*) to the country of origin.

Conclusion

This first part of the book has sought to conceptualize the phenomenon of climate change-related migrations in a manner that is relevant and useful to an international law analysis. In this regard, we have seen first why considering anthropogenic climate change as the primary or exclusive cause of numerically exorbitant forced displacements and international migrations worldwide is misleading. In reality, the choice to migrate depends on a complex set of factors, among which global warming and environmental pressures can be important but rarely exclusive or decisive.

[40] See in this sense van der Vliet (2018). "Climate Refugees", a Legal Mapping Exercise. In Behrman & Kent (Eds.) (op. cit.), p. 23.

[41] Shacknove (1985). Who Is a Refugee? *Ethics*, 95, p. 275.

In this logic, it is sounder to conceive global warming as a "threat multiplier", i.e., as a factor which can exacerbate already socially, economically and politically fragile situations in some areas of the world. It follows that the relationship between climate change and migration largely depends on the context in which a given climate-induced environmental degradations occurs: political, economic, and institutional pre-conditions have significantly profound effects on the degree of resilience of a community and, likewise, on its vulnerability. Even when a direct link between climate change and a progressive or immediate environmental disaster exists, human migrations linked to it remain multi-causal. Put simply, it is the interaction between the *ensemble* of effects caused by climate change and the pre-existing conditions on which these impact that will eventually determine, if, why, and when human beings leave their home to seek better opportunities elsewhere.

This first part also stressed that the distinction between climate-related voluntary movements and forced displacements (i.e., where the choice to stay is precluded) has to be seriously taken into account. Although such a distinction is not always easily identifiable, it is decisive for an international law analysis. More precisely, it is crucial when it comes to international protection. In fact, only those who we defined as "forced climate migrants"—i.e., individuals compelled to leave their origin country because its territory has become (or is about to become) uninhabitable due to climate change-driven environmental degradations—present the characteristics by virtue of which a third state should be obliged to provide them with protection and assistance. On the other hand, individuals migrating (more or less) voluntarily *also* because of climate change pressures remain able to enjoy a relationship of trust and protection, albeit limited, with their national country. These people are thus not in the immediate and acute need for assistance in a third state, which in turn should not be responsible for their protection.

Thanks to this approach, the group of those for whom international law should take action with a view to guaranteeing some form of protection abroad shrinks significantly. We saw that the adoption of excessively general and all-encompassing conceptualizations trying to define climate-induced migrants has led to the creation of what we called the "climate refugees cauldron": that confused set of individuals, communities and populations which have been merged together and considered, without any distinction, in potential need for international protection because they have left their homes—be it voluntarily or forcibly, internally or internationally—directly or indirectly in presence of climate-related environmental pressures. This book argues however that with the view of assuring protection standards to the most vulnerable individuals, it is easier and more feasible to focus on a specific subset of this “cauldron”, namely, on those who must leave the territory of their state due to environmental degradations caused by anthropogenic global warming, and we called these persons “forced climate migrants”. In this sense, the group of people that best epitomizes this reality is, without a doubt, that of the inhabitants of “sinking” micro-state islands which risk becoming completely “deterritorialized” because of sea-level rise over the next few decades.

PART II

A Problem of Rights and Responsibilities: The Legal Rationale for the Protection of Forced Climate Migrants

Introduction

One of the Gordian knots concerning climate change-related migration arises from one simple but crucial question: why? Why should people forced to migrate due to climate change be protected abroad? Too often, in fact, we take the answer for granted: global warming is the work of mankind, it is therefore right that those who are most vulnerable to its effects and are forced to leave their countries as a result, must be protected in other states. This reasoning has logical value, but it is still not supported by a comprehensive analysis of the international obligations existing today in the field of human rights and climate change law. In simple words, if we cannot identify the reasons why and the legal bases by virtue of which the category of forced climate migrants should be protected under international law, we risk studying a subject that should not be looked at through the lenses of international protection systems. Without a specific legal basis, it is unsound to talk of protection: with the worsening of global warming and its associated environmental degradations, there will "just" be an increase in "economic" migratory flows, not "worthy" of activating the *non-refoulement* principle, understood as the "prohibition against sending someone back to where they would face torture or other cruel, inhuman or degrading treatment or punishment, or threat to their right to life, which is termed 'complementary protection' under international law".[1]

This second part of the book studies existing international law principles, and rules to understand the extent to which forced climate migrants (as we defined them in Part I) may be entitled to international protection. It identifies the legal reasoning by virtue of which international law can, as was in past centuries the case for "political refugees", provide tools to protect a historically new category of forced migrants: "climate refugees". Supporting the existence of an international duty to protect forced climate migrants may be morally trivial, but it is actually quite complex from a legal viewpoint. Hence the need to investigate the problem through a systemic or holistic approach that takes into account state responsibility rules, the origin of international customs and *erga omnes* obligations, and the relationship between anthropogenic climate change and human rights law.

The reasoning we propose here moves from the general to the particular: starting from a general assessment of existing state obligations to mitigate and manage global warming, it enquires whether among these a particular obligation to protect forced climate migrants is to be put in place. The main aim is to understand if and how the issue of forced climate migrations can be addressed through a climate justice and human rights perspective, not in the sense of attributing the responsibility to protect forced climate migrants to the polluting states, but rather to investigate the international community's opportunities to create and enact protection systems for forced climate migrants. Of course, states enjoy, generally speaking, broad flexibility in regulating the movement of foreign nationals across their borders. They have the right to exercise their sovereignty and to decide who can be admitted and for what period within their territory, but this authority is not absolute.[2]

In the following pages, we will try to understand if there are international law principles by virtue of which a state should be obliged to

[1] See Poon (2018). Drawing upon International Refugee Law. The Precautionary Approach to Protecting Climate Change-Displaced Persons, in Behrman & Kent (Eds.) (2018). *'Climate refugees' Beyond the Legal Impasse?* New York: Routledge, p. 160.

[2] Martin (2017). Toward an extension of complementary protection? In Mayer & Crépeau (Eds.). *Research Handbook on Climate Change, Migration, and the Law*. Cheltenham: Elgar, p. 449.

assist and protect a foreign citizen fleeing his or her homeland due to climate change. We are going to look at universally recognized human rights whose deprivations may be caused by global warming, leading individuals to forcibly leave their country. Complementarily, we will look at those obligations, existing in the name of the protection of these rights, which fall upon the international community as a whole and its states. The basic idea is not yet, therefore, to show how forced climate migrants can or should be protected under existing instruments, but rather to identify the reason why we can talk about the slow-onset birth of an international duty to create and enact protection systems for this category of migrants. To this end, a thorough study of states' international obligations and responsibilities *vis-à-vis* climate change and human rights is needed.

When we discuss the notion of responsibility here, the aim is not to focus merely on what we may call "canonical" state responsibility. Responsibility is conceived more broadly here. Thus, the focus is not only on the classic notion of international responsibility, one that arises from both licit and illicit acts carried out by a state and that have negative effects on the sovereign rights of other states. Such an interpretation is obviously taken into consideration, especially when it comes to climate change mitigation and adaptation policies that might be undertaken mandatorily both under general and particular international law. However, the attempt here is to reason on a broader understanding of the concept of responsibility. To this end, the analysis will focus on three different dimensions of the concept. First, we try to understand what international obligations exist today in relation to climate policies that should be undertaken under customary law, thus through the "do no harm" principle, and treaty law, namely under the UNFCCC and its implementation protocols (Kyoto and Paris). Second, the following chapter investigates whether different elements arising from this set of rules dedicated to climate change may shed light on a new international custom under which states would be required to cooperate and act in the fight against and in the management of global warming. Given that this is the case, Chapter 7 will argue that these obligations would consist in *erga omnes* obligations, and this is because of the essential importance of the international "good" that they aim to protect: the climatic balance on which the international community has lived, thrived, and developed for several centuries. Third, and more specifically with reference to the issue of climate change-related migration, we will investigate state obligations regarding human rights: we shall see that the enjoyment of some inalienable fundamental rights can be put at serious risk because of the negative effects of climate change. In this context, the discussion on state obligations (both with an internal and extraterritorial scope) is developed with the aim of understanding whether states should undertake climate mitigation and adaptation policies—among which the development of protection mechanisms for the protection of forced climate migrants—in the name of the protection of these non-derogable rights.

Therefore, the main aim is not so much to determine whether and to what extent a given country can be held responsible for a particular "wrongful climate act" under which it would, for example, be required to host and protect certain quotas of "climate refugees" as a form of reparation for the damages caused.[3] The goal is rather to identify the legal bases on which we may argue that—given the relationship between certain economic activities which are promoted and permitted by states and their negative impact on human rights and forced migration—international law should equip itself in the coming years, as it did over the past centuries for "political refugees", with instruments designed to protect a new category of forced migrants: "forced climate migrants".

[3] See in this regard for instance: Byravan & Rajan (2006). Providing New Homes for Climate Change Exiles *Climate Policy*, 6, p. 247 ff. Some studies, such as this one, focus on the fact that "climate refugees" should be admitted to foreign territories according to the quantity of emissions produced by each single state. However, such an approach probably does not sufficiently take into account of all the legal problems related to the issue of identifying international responsibility for climate change. Furthermore, the proposers also risk distancing themselves from the protection of the human and collective rights of the people involved. Indeed, maintaining that "climate refugees" should be transferred from one country to another in the name of the political and ethical responsibility of the most polluting states could relegate those affected to the rank of "exchange goods". If solutions of this kind were adopted, the guarantee of the cultural and collective rights of the displaced communities—and also those of destination—would be severely tested. As shown, migrations understood as forms of adaptation to changes—environmental and otherwise are part of the culture of many communities. It is therefore necessary to study existing migratory flows and try to make these legal, effective, and safe, in order to respect the rights and values of the communities concerned, rather than create new corridors, moreover on a more or less coactive basis. Such an approach ultimately risks complicating the discussion, rather than providing an intelligent way out.

CHAPTER 6

International Climate Change Law and State Responsibility

When we study the issue of state responsibilities *vis-à-vis* climate change, it is first necessary to focus on what we may call "canonical" international responsibility. The purpose of this chapter is to start understanding to what extent the (im)possibility of exercising certain fundamental rights in some national legal systems (which may lead to forced migrations) can be linked to activities that are somehow attributable to the subjects of international law.

In matters of state responsibility, the distinction between "responsibility" and "liability" is well known among international legal scholars. While state liability applies to lawful acts that have produced, despite their lawfulness, damage to another country, state responsibility applies in reaction to acts that are illegal in their very nature.[1] In this sense, and making reference to Latin, it is possible to distinguish between responsibility *sine delicto* and responsibility *ex delicto* (with and without delict). When climate change and its intrinsic complexity come into play, both notions

[1] Cf. Hoss (2005). State Responsibility, Liability, and Environmental Protection. In Wolfrum, Langenfeld & Minnerop (Eds.), *Environmental Liability in International Law: Towards a Coherent Conception*. Berlin: Erich Schmidt Verlag, pp. 455 ff. See also: Kiss, Shelton (2007). Strict Liability in International Environmental Law. In Nndiaye & Wolfrum (Eds.), *Law of the Sea, Environmental Law and Settlement of Disputes: Liber Amicorum Judge Thomas A. Mensah*. Leiden: Martinus Nijhoff Publishers, pp. 1131 ff.

G. Sciaccaluga, *International Law and the Protection of "Climate Refugees"*,
https://doi.org/10.1007/978-3-030-52402-9_6

are relevant.[2] Greenhouse gas emissions are normally and up to now overwhelmingly considered as acceptable and completely lawful "acts". They can nonetheless cause damages in third countries when they start altering global climate patterns. Under this perspective, greenhouse gas emissions can become part of the set of *sine delicto* responsibilities. At the same time, however, a historically circumscribed part of greenhouse gas emissions can potentially also take the form of illicit acts. When this happens, the economic activities generating such emissions could then be considered as part of the realm of *ex delicto* responsibilities. This would be the case when a state fails to undertake measures to mitigate its own emissions even though it is under an obligation to do so, be it due to customary international environmental law or to treaty law dedicated to climate change and, therefore, from the UNFCCC and its implementation instruments (the 1997 Kyoto Protocol and the 2015 Paris Agreement).[3]

To better understand state responsibility *ex delicto*, the 2001 International Law Commission's Draft Articles on Responsibility of States for Internationally Wrongful Acts are highly relevant.[4] By codifying existing customary rules, the Draft clarifies that a state is to be considered responsible for a particular conduct (that can also be of omissive nature), if this is attributable to the state ("subjective element") and, at the same time, if it violates an international obligation ("objective element").[5] Therefore, a

[2] Koslopova (2013). *Interstate Liability for Climate Change-Related Damage*. The Hague: Eleven.

[3] For sake of completeness, it is also true that some conventional obligations dealing with greenhouse gases can also be found outside the UNFCCC system. For instance, in the Montreal Protocol, whose objective is to limit the anthropogenic depletion of the Ozone Layer. By limiting chlorofluorocarbons (CFC), which as a gas have indeed a climate-altering potential, the Montreal Protocol marginally addresses by its very nature global warming causes. The 2016 Kigali Amendment on the Protocol then intensifies the latter's role in the subject matter by also imposing reductions o hydrofluorocarbon (HFC) emissions, which again are powerful greenhouse gases. Besides the Ozone Layer regulatory system, there are also attempts to include climate change mitigation objective within aviation and sea-related international conventions. The 1973 Convention for the Prevention of Pollution from Ships (so-called MARPOL Convention), now aims at including regulations to limit GHG emissions from ships. An International Civil Aviation Organization (ICAO) resolution of 2016 has started to lay out market-oriented schemes for the reduction of GHG emissions from the aviation sector applicable from 2027.

[4] ILC Draft Articles on Responsibility of States for Internationally Wrongful Acts, attached to UN Doc. A/Res/56/83, 12 December 2001.

[5] On this subject matter, see *ex multis*: Fumagalli (2016). Illecito e Responsabilità. In Carbone et al. (Eds.), *Istituzioni di Diritto Internazionale*. Turin: Giapichelli, pp. 269 ff.

state could hypothetically be held responsible for a climate change-related forced migration if a conduct attributable to it and prohibited under international law is the cause of the specific migratory event. However, it is extremely difficult (or, better, impossible) to clearly establish this causal link given the nature of anthropogenic climate change and of human migration itself, which, as discussed in Chapter 3, are two exceptionally complex fields of study. Anthropogenic climate change is a phenomenon that is in fact attributable to human activities, but ones that have taken place—if we accept the date arbitrarily chosen by the IPCC with respect to the beginning of the industrial era—starting from 1850, and for which a more or less easily identifiable group of states, "industrialized ones", is primarily and collectively responsible. Huge scientific uncertainties and intergenerational concerns make the direct link between some given greenhouse gas emission and a given migration anything but unidentifiable. It is obvious then that among the cornerstones of international climate change law we find the principle of common but differentiated responsibilities: global warming is a phenomenon with global and widespread origins, and the degree of responsibility for its manifestation varies according to the emissions generated by each single state over the course of their entire industrial history, spanning over a period of almost two centuries.[6]

It is in this sense that the identification of a precise causal link between the activity of a certain state and any damage caused in others due to climate change (and, at a further level, also the link with a consequent forced migration of individuals), and hence the identification of the aforementioned "subjective element", becomes not only complicated but also clearly impossible. There are simply too many variables at stake. It is nevertheless possible to argue that the international community as a whole (and within it some states particularly) can be held responsible for climate change and its associated damages, especially when states have

See also: Crawford et al. (2010). *The Law of International Responsibility*. Oxford: Oxford University Press.

[6] See UNFCCC, Preamble, para. 3 and 6: "Noting that the largest share of historical and current global emissions of greenhouse gases has originated in developed countries, that per capita emissions in developing countries are still relatively low and that the share of global emissions originating in developing countries will grow to meet their social and development needs"; "acknowledging that the global nature of climate change calls for the widest possible cooperation by all countries and their participation in an effective and appropriate international response, in accordance with their common but differentiated responsibilities and respective capabilities and their social and economic conditions".

not seriously engaged in greenhouse gas emissions curbing policies even after the awareness of their negative effects has arisen worldwide. Hence, even if it is difficult or even impossible to identify the legal responsibility of a particular state, the existence of a "collective" responsibility of the heavy polluting countries and, beyond these, of the whole international community is clear.[7]

Similar complications also exist with regard to the identification of the "objective element". Understanding whether greenhouse gas emissions are (un)lawful acts is not easy at all. It is necessary to clarify how and on the basis of which existing legal dispositions GHG emissions that are not adequately capped by a state can be considered as internationally wrongful acts, and this is again euphemistically difficult. In the first place, greenhouse gas emissions are (if only for most of their historical existence) a completely accepted and lawful activity, not to say promoted by the (inter)national political-legal systems with a view to enhancing economic growth and social development.[8] Second, if it is true that state-owned entities or companies can directly emit greenhouse gases, it is equally true that private actors (industries, transport, families, and, so on) generate their vast majority. In fact, the private sector is the main global GHG emitter, and this is surely true in market-oriented liberal democracies. In this context, a state can be accountable for the conduct of private actors acting under its jurisdiction only if and when its public authorities have failed in controlling, limiting, or stopping such conduct despite having an obligation to do so in order to avoid possible damage arising within their own borders or in other states' territories.

This reasoning, which in international environmental law gives birth to the "do no harm principle", is part of customary international law,

[7] "[It is possible] to establish that the law of State responsibility is applicable to at least some States, either because they fail to fulfill their mitigation commitment under climate change agreements or because their excessive greenhouse gas emissions cause disproportionate harms to the global atmospheric system, in breach of the no harm/preventive principle". In Mayer (2017). Climate Change, Migration and the Law of State Responsibility. In Mayer & Crépeau (Eds.), *Research Handbook on Climate Change, Migration and the Law*. Cheltenham: Elgar, p. 255.

[8] In this sense see also: "Necessary anthropogenic greenhouse gas emissions include most obviously those resulting from human breathing but also, arguably, the emissions that, in a particular technological and development context, are inevitably generated in order to reach a minimal level of human development. It would concededly be extremely difficult to determine how much greenhouse gas emissions are thus justified as necessary and from which threshold emissions would become excessive". Ibid., p. 248.

and requires states to refrain from—and not allow—a use of their territory capable of causing significant environmental damage beyond their national borders, in accordance with the old maxim "*sic utere tuo ut aliaenum non laedas*" from which the "do no harm" principle derives.[9]

States have a negative obligation not to cause significant environmental harm in other countries (do no harm principle) and a positive obligation (due diligence) to prevent that activities carried out within their jurisdictions may cause such cross-border harm. The *sic utere tuo principle* is broadly recognized as the cornerstone of international environmental law and as part of international customary law, as explicitly pointed out by the International Court of Justice in 1996 for instance.[10] Moreover, both the 1972 Stockholm and 1992 Rio Declarations on environmental matters have proclaimed that states are to follow this principle.[11] Under its application, there is an intrinsic conflict between the right of each state to allow in its territory activities capable of producing damage from pollution, and the right of every other state to prevent such damages from also taking place in their own territory. From a theoretical point of view, the do no harm rule originates from the concept of abuse of right and/or from the category of good neighborly obligations understood in a broad sense.[12] A state can be responsible (and this is relevant both from the standpoint of responsibility *ex delicto* and *sine delicto*) where no action is taken to prevent the occurrence of foreseen and foreseeable damages in other states. Therefore, *foreseeability* plays a pivotal role here.

[9] The do the International Court of Justice has referred to no harm principle. See for instance: *Corfu Channel Case (UK c. Albania)*, 1949, I.C.J Rep 4, p. 22. See also *Trail Smelter Arbitration* (USA v. Canada), where the Tribunal concluded that: "under the principles of international law, as well as of the law of the United States, no State has the right to use or permit the use of its territory in such a manner as to cause injury by fumes in or to the territory of another or the properties or persons therein, when the case is of serious consequence and the injury is established by clear and convincing evidence".

[10] See International Court of Justice, *Legality of the Threat or Use of Nuclear Weapons, Advisory Opinion*, 1. C.J. Reports 1996, p. 226, para. 242.

[11] See the 1972 Stockholm Declaration, Principle 21. 1992 Rio Declaration, Principle 21. In legal literature, see *ex multis*: Bratspies & Miller (2006). *Transboundary Harm in International Law: Lessons from the Trail Smelter Arbitration*. Cambridge: Cambridge University Press.

[12] Cf. Picone (2013). *Comunità Internazionale ed Obblighi Erga Omnes*. Naples: Jovene Editore, p. 13.

Climate change surely remains a relatively recent phenomenon and a "super-wicked" problem, but it is also true that the first IPCC report dates back to 1990, while the entry into force of the UNFCCC, a treaty that enjoys 197 ratifications and in which—sobering to remind—the human origin of global warming is ascertained,[13] dates back to 1994, twenty-six years ago. The UNFCCC imposes a general obligation for developed state parties to "*adopt nationall policies and take corresponding measures on the mitigation of climate change, by limiting its anthropogenic emissions of greenhouse gases and protecting and enhancing its greenhouse gas sinks and reservoirs*",[14] and for developing ones to "*formulate, implement, publish and regularly update national and, where appropriate, regional programmes containing measures to mitigate climate change by addressing anthropogenic emissions*".[15] Clearly, the UNFCCC follows a soft and general approach with regard to states obligation *vis-à-vis* climate change. Indeed, this reflects its nature as a framework convention. Through this approach, the UNFCCC "boils down to little more" than a general obligation to do something about global warming, and it does not request to do anything in particular.[16]

This is without question, but the now-dated entry into force of the UNFCCC shows that states have had quite a long period of time to adapt to anthropogenic global warming and, above all, to understand its associated threats. When bearing this in mind, it is not fanciful to argue that those states that since then have not undertaken acts of law and policies to progressively curb their greenhouse gas emissions may be held in violation of the do no harm principle, and this (if nothing else) especially with reference to the interests and rights of insular micro-states at risk of submersion and deterritorialization.

Reasoning with a view to *ex delicto* responsibility, this means that in application of the do no harm principle, no GHG emission should and could be considered wrongful if it dates back to before 1990, or—in application of the obligations arising *ex* UNFCCC—before the entry into force (and only for the states that have adhered to it) of the Kyoto Protocol, the first agreement to explicitly introduce emission reduction

[13] See UNFCCC, Preamble and Articles 1 and 2.

[14] UNFCCC, Article 4(2)(a).

[15] UNFCCC, Article 4(1)(a).

[16] See Mayer (2017) (op. cit.), p. 243.

obligations, establishing objectives to be reached in the years by developed and industrialized states, first for the period 2008–2012, and then, through the Doha Amendment—that however did not enter into force—also for the 2013–2020 period. So, in the end, only a very small part of historical GHG emissions is effectively covered by legal limits, both under customary and/or treaty law.[17] This does not entail, however, that through an extensive interpretation of the do no harm rule, large greenhouse gas emitters cannot be held liable for climate-related damages or right infringements occurring in other countries where they have increased (or not adequately and proportionally limited) their emissions, at least after joining the UNFCCC.

Of course, this reasoning has its evidently weak spots. By the beginning of the Nineties, when the Framework Convention came to light, the scientific and, consequently, also the political and legal certainties regarding global warming were relatively fragile, allowing wide margins of appreciation to states as to what they should or should not have done to address climate change. However, if there can be reasonable doubt regarding the existence and effectiveness of obligations and responsibilities in the fight against climate change with respect to the 1990–2015 period—a period that at a legal-historical level, and despite the clarity of the terms used in the UNFCCC, can be recalled as the "prelude" to the establishment of an incisive and widely shared international climate change regime—such doubts have significantly reduced following the entry into force of the Paris Agreement in 2015.

While it is not among the aims of this work to dwell on the evolution of international climate change regime, the Paris Agreement can nonetheless be considered as the "year zero" of a new era, since it puts in place a normative architecture potentially capable of lasting several decades.[18]

[17] It is sobering to remember that international responsibility rules cannot be applied retroactively. Cf. ILC Draft Articles on Responsibility of States for Internationally Wrongful Acts, attached to UN Doc. A/Res/56/83, 12 December 2001, Article 13.

[18] See: Bodansky (2015). Reflections on the Paris Conference, in *Opinio Juris*. http://opiniojuris.org/2015/12/15/reflections-on-the-paris-conference/. Accessed 14 January 2019. Doelle (2015). The Paris Agreement: Historic Breakthrough or High Stakes Experiment? https://papers.ssrn.com/sol3/papers.cfm?abstract_id=2708148. Accessed 14 January 2018. European Commission (2016). The Road from Paris: Assessing the Implications of the Paris Agreement and Accompanying the Proposal for a Council Decision on the Signing, on Behalf of the European Union, of the Paris Agreement Adopted Under the United Nations Framework Convention on Climate Change. Communication

Above all, however, it appears sound to talk of a new era because the Agreement is based on the shared conviction that every single state now has an obligation to act in response to the activities that cause global warming.[19] In the recent past, the possibility to avoid climate change obligations appeared more available than today. Interestingly, Burkett reasons as follows: "in recent history, there was a more reasonable alternative conduct under the circumstances. In other words, a less carbon intensive economy has been possible for decades, a political economy for which there was significant advocacy. [...].Climate change is the result of a very particular way of doing business. It is the massive externality of a chosen political economy".[20]

Today, with the ever-growing evidence that our international political economy is leading to very pressing international challenges and human suffering, the need to do something about it has commensurately grown. The Paris Agreement, with its five-year ratchet-up structure clearly shows

(COM(2016) 110 final). Huang (2015). The 2015 Climate Agreement: Key Lessons Learned and Legal Issues on the Road to Paris. https://papers.ssrn.com/sol3/papers.cfm?abstract_id=2724109. Accessed 14 April 2019. Baker & McKenzie (2016). The Paris Agreement Putting the first universal climate change treaty in context. https://www.lexology.com/library/detail.aspx?g=7f89d079-a97c-4c87-99ba-16f3c9ff5fdf. Accessed 13 January 2018. Obergassel et al. (2016). Phoenix from Ashes—An Analysis of the Paris Agreement to the United Nations Framework, Wuppertal Institute for Climate, Environment and Energy. https://epub.wupperinst.org/frontdoor/deliver/index/docId/6374/file/6374_Obergassel.pdf. Accessed 13 January 2018. Gervasi (2016). Rilievi critici sull'Accordo di Parigi: le sue potenzialità e il suo ruolo nell'evoluzione dell'azione internazionale di contrasto al cambiamento climatico, *La Comunità Internazionale*, p. 21 ff. Romanin Jacur (2016). The Paris Agreement on Climate Change: Progresses Achieved and Challenges Ahead. *NCTM Newsletter*. https://www.nctm.it/news/articoli/the-paris-agreement-on-climate-change-progresses-achieved-and-challenges-ahead. Accessed 12 January 2018.

[19] With an opposite view, Mayer does not understand the whole evolution of international climate change law as the states' determination to assume rights and obligations vis-à-vis the management of the atmosphere: "Every climate change agreement has embodied a notion of progression towards more demanding obligations until the ultimate objective of preventing a dangerous anthropogenic interference with the climate would be reached. Therefore, climate change agreements can be understood as a transitory regime of collective emulation and collaboration, but not as the definitive determination of States' rights and obligations with regard to their usage of the atmosphere" in Mayer, Climate Change, migration and the law of State responsibility, in Mayer (2017) (op. cit.), p. 244.

[20] Burkett (2018). Justice and Climate Migration, The Importance of Nomenclature in the Discourse on Twenty-First-Century Mobility. In Behrman and Kent (Eds.) (op. cit.), pp. 73 ff.

that the political and legal uncertainties regarding the necessity to curb national and global greenhouse gas emissions have less and less room in the international community's reasoning. Of course, the Agreement's strength and effectiveness still needs to be tested. Only with a widespread support by the leading industrial countries, there is a chance for it to be successful. Undeniably, the "schizophrenic" attitude of the United States with regard to global warming plays a pivotal role in all of this. Without a deep commitment of the most economically, militarily, and technologically developed country in the world on the need to seriously curb greenhouse gas emissions, the chances to cope with climate change in socio-economically sustainable ways at a global level shrink significantly. Nonetheless, the position of the American government, whose withdrawal from the Paris Agreement will take effect by the end of 2020,[21] while generating enormous uncertainty given the weight of the United States in world affairs, does not entail lack of commitment of the international community as a whole to seriously tackle climate change. The adventure undertaken by President Trump remains, in fact, quite solitary, with no other country in the world backing it and, indeed, with the most significant actors in terms of GHG emissions worldwide (such as China, EU, Russia, Japan, Canada, India) having harshly criticized it. Moreover, the double layer of American politics allows key economic federate states, such as California for instance (that would be the world's fifth largest economy if it were independent), not to back federal policy on climate change. In fact, many American states, local authorities, and private companies have decided to back the Paris Agreement and pursue its objectives notwithstanding the position of the White House.

As a whole, the Paris Agreement can be seen to a certain extent as a historical milestone because of its structure and design, and, at a deeper and more important level, because it is fundamentally based on the shared awareness that every and any international law subject must do its part to address the "biggest crisis humanity has ever faced", as (not only) the teenage climate activist Greta Thunberg has put it.

As we mentioned, the Agreement puts in place a ratchet-up normative structure that can potentially be effective for decades. It has set out a shared and common global mitigation goal: keeping the increase in the average terrestrial temperature beneath 2°C, preferably 1.5°C, to assure

[21] Paris Agreement, Article 28.

the long-term sustainability of ecological and socio-economic balances on the planet. This common goal is to be achieved through voluntary efforts of the single parties, that collectively have to reach the global (and hence their national) peaking of greenhouse gas emissions as soon as possible. More specifically, each state is required to submit its national GHG reduction contribution programs, the so-called nationally determined contributions (NDC) to the UNFCCC Secretariat every five years. NDCs are self-imposed ever more ambitious five-year programs aimed at progressively curbing national greenhouse gas emissions. The most interesting aspect in all of this is that the Agreement imposes that parties must proceed in this direction with no time limit, meaning that they bear an obligation to submit and implement NDCs that are ever more ambitious literally for decades. "*Every Party shall prepare, communicate and maintain successive nationally determined contributions that it intends to achieve [and] each Party's successive nationally determined contribution will represent a progression beyond the Party's then current nationally determined contribution [...]*".[22]

Through this ratchet-up mechanism, national, regional, and eventually global climate policies are to become increasingly "aggressive" and effective over time, hence (theoretically) assuring in the long run the socio-economic and geopolitical sustainability of climate change. Alongside this, the Agreement establishes that every five years starting from 2023 a "global stock-take" must be put in place, which is an overall analysis and assessment on the effective implementation of every party's NDC throughout time. As often the case in international law, the system's weakness lies, as broadly highlighted at every level, in its lack of enforcement and sanctioning measures in case a state fails to fulfill its NDC's objective. The system is based, in fact, on the principles of voluntariness and international transparency and, ultimately, on the shared awareness that a common but differentiated worldwide effort is needed to meet the 2°C objective by the end of this century. Put simply, the Paris Agreement is based on (and can only be successful thanks to) the conviction that "we are all in this together", and that we all must act to prevent the worst climate change scenarios.

Through its successive and ever more ambitious five-year plans, the Agreement's architecture is aimed to last long. It remains to be seen,

[22] Paris Agreement, Articles 4.2 and 4.3.

of course, how and especially how much states will be willing and able to undertake the emissions' cutting policies submitted in their NDC. As of today, apart from the significant but solitary US position, international support for the Paris Agreement is strong and widespread (at least rhetorically and symbolically), with as many as 189 ratifications and the presentation of at least 186 NDCs. This overarching support suggests that, despite its inevitable flaws and weaknesses, the normative system sets out with the Agreement could prove to be a successful international reference model in the coming years. And this is contrary to what the Kyoto Protocol represented. The latter, in fact, imposed top-down obligations to cut national greenhouse gas emissions at a time when the common intent to act against climate change was decidedly much lower than today. This approach proved to be unsuitable to inspire a determined action by the vast majority of states,[23] not to talk about the fact that huge developing countries remained completely free to pursue their economic development without paying attention to their contribution to global warming. After more than twenty years, the Kyoto Protocol can be undoubtedly looked at as a historic failure, covering only about 12% of global greenhouse emissions and with only a very small number of countries (European Union Member States alongside with Switzerland, Iceland, and Norway) having accepted to undertake emissions cut commitments up to 2020. This explains why the normative approach has completely changed with the Paris Agreement, which, allowing for a much greater flexibility, tries to co-opt as many actors as possible.

Coming to a conclusion, global warming surely remains a historical novelty, something that is moreover intrinsically complex. And international law moves forward slowly, probably too slowly when it comes to acknowledging state responsibilities. No surprise then that climate change international law is not capable of adequately addressing the problem of international responsibility. Both customary and treaty law fail in setting clear criteria thanks to which it would be possible to identify those legal subjects which are responsible for global warming as well as the reparations of the damages they cause. As a consequence, from a legal-historical perspective, the approach in fighting and managing climate change has

[23] As known, the United States refused to comply with the system and did not ratify the Kyoto Protocol. Furthermore, Canada and Japan have withdrawn from it before the end of the first commitment period, leaving in the end just the EU Member States, alongside with Switzerland, Iceland, and Norway to comply with it.

shifted from imposing top-down goals to industrialized states, to setting a common and global objective that is to be reached through to the effort of the entire international community, no one excluded. The nature and language of the Paris Agreement clearly show that the focus today is on voluntary efforts by every party, to be made according to each state's capability and historical responsibility for greenhouse gas emissions. States remain free to choose their policies and objectives, with enormous flexibility allowed. No superior authority imposes anything: it is the common and shared sense of responsibility that should lead every state to do its best to address climate change, mitigate it, and adapt to it. The principles of voluntariness and transparency are crucial. The whole Paris Agreement's system can only work through a serious and open international cooperation effort, with state parties reciprocally monitoring each other's behaviors, thus creating peer pressure to constantly and progressively move collectively toward the 2°C objective. If there is a widespread awareness on the nature and effects of climate change and a strong conviction to act collectively to avoid its direst effects, the whole system might work, if not, it will fail. In the absence of international enforcement mechanisms, such collective awareness represents the only pathway that, while long and fraught with hurdles, has a chance to deliver concrete change.

The ever-growing amount of scientific evidence on the need to seriously address the problem might lead to a shared level of awareness and conviction across the international community. Let us just hope this happens before it is too late to avert irreversible climate consequences.

At the same time, this growing awareness leaves less and less room to put aside international customary law and the *sic utere tuo* principle. The do no harm principle requests states not to provoke transboundary damage caused by activities they carry out in their own territories, also imposing upon them the obligation to limit and control private actors that operate under their jurisdictions. It is nowadays increasingly difficult, and with the ongoing research in climatology it will more and more so, to argue that a failure in limiting national greenhouse gas emissions does not cause any damage abroad. But, of course, climate change is a "super-wicked" problem, and it will always be impossible to determine which particular emission provoked which particular damage abroad. There will always be someone "*à la* South Park's Eric Cartman" out there arguing that he or she cannot be held responsible for any particular damage caused throughout the world by global warming. This is obviously true, since no individual subject, be it a person, a business company, or a state can be held responsible for any particular damage climate change causes. The truth is that we are all in this together, and that it is not easy at all to envisage and create legal obligations in such a context. Both customary and treaty law, however, request today that something is to be done to address climate change. Following this trend and this historical evolution of climate change international law, arguably those states that persevere in ignoring the problem and in failing to regulate and limit their national greenhouse gas emissions should be held responsible for a breach of international law.

In conclusion, the mitigation of climate change, the management of its effects, and the adaptation to them have now become an integral part of the international legal system and of the international community's objectives. Even though this evolution occurs slowly, maybe too slowly to avert desolating future scenarios, it tends toward establishing obligations on every state to do something about the problem. Furthermore, if we consider that the very survival of the international community may depend on the success of current and future international climate policies, a further legal argument arises to strengthen the idea that states are requested to act swiftly. If we depend upon the protection of our climate system for the functioning of the entire international community, then the action needed to protect it derive from *erga omnes* obligations, something that puts them in a very particular position.

CHAPTER 7

The *erga omnes* Obligation to Mitigate and Manage Climate Change

As we outlined in the previous chapter, the fight against climate change and the management of its negative effects have now become an integral part of the objectives and obligations of the international community. Building on this, this chapter focuses on the nature of these obligations, arguing that, given the essential importance of the common good that they aim to protect, i.e., the stability of the climate system, they stand as *erga omnes* obligations.

If this is true, once the human origin and the potential uncontrollable effects of climate change are agreed and ascertained, every state, although with differentiated responsibilities, would be obliged to fight and manage global warming through mitigation and adaptation measures. Thus, with reference only to one specific dimension of the challenge we are starting to face, we may argue that within these measures, states would also have the responsibility to provide international protection to forced climate migrants. To support this argument, a careful examination of the role and nature of *erga omnes* obligations and their relationship with climate change is needed.

Under international law, an *erga omnes* obligation arises where a particular phenomenon, because of its intrinsic nature and importance,

G. Sciaccaluga, *International Law and the Protection of "Climate Refugees"*,
https://doi.org/10.1007/978-3-030-52402-9_7

must be managed by and on behalf of the entire international community or, better said, by all states operating *uti universi*, or collectively.[1] From this viewpoint, *erga omnes* obligations are not merely "horizontal", in the sense that they reciprocally apply between states understood as equal, sovereign, and independent subjects. They are rather "vertical" obligations: states bear an obligation that is to be fulfilled on behalf of the interests of the entire international community, understood as a stand-alone subject.

This peculiar kind of obligation comes at play when a primary and essential transnational good or value must be safeguarded, "something" on which the whole international community bases its existence and its functioning, and whose protection can only be assured by all states if they operate collectively.[2] To protect these goods and values, which by their very nature reflect global public goods and common interests (in fact, we may here refer to the common heritage of humanity), every single state bears a behavioral obligation that can be demanded, at least hypothetically, from the collectivity of the other states, which operate on behalf of the entire international community. This happens because similar essential values, precisely because they "belong" to the international community and not to its single components, are "unavailable" to the individual state.

The evolution of international law over the last centuries can be understood in this sense: from a law of mere "interstate coexistence", international law has progressively evolved into the law of "interstate cooperation", to finally become a system that regulates "interstate interdependence". Global challenges and phenomena, such as increasing international economic integration, the spread of pollution, and the severe depletion of natural resources on a transnational scale, require an international legal system that is capable of managing, in the name and on

[1] For a detailed analysis of the nature and characteristics of *erga omnes* obligations, please refer to Picone (2013). *Comunità Internazionale ed Obblighi Erga Omnes*. Naples: Jovene Editore, Chapter 1.

[2] The first hint regarding *erga omnes* obligations is to be found in a famous *obiter dictum* by the International Court of Justice. See *Barcelona Traction* (ICJ, *Barcelona Traction, Light and Power Company, Limited*, I.C.J. Reports, 5 February 1970, p. 3), para. 33: "In particular, an essential distinction should be drawn between the obligations of a State towards the international community as a whole, and those arising vis-à-vis another State in the field of diplomatic protection. By their very nature the former are the concern of all States. In view of the importance of the rights involved, all States can be held to have a legal interest in their protection; they are obligations *erga omnes*".

behalf of the community of states, those collective values on which the functioning of the international community ultimately depends.[3] The evolution of the international system in this direction, although clearly anchored to the notion of state sovereignty, is clear. The effectiveness of the system it brought about can however be questioned.

The work carried out by the International Law Commission (ILC) on the aforementioned Draft Articles on State Responsibility for Internationally Wrongful Acts, which began back in 1976, is highly relevant in this subject matter. Although the last Draft, which was adopted in 2001, does not make reference to this point, the precedent ones opted for a neat distinction between international "crimes"—that are committed against the international community—and "delicts"—committed against the single state. While crimes affect public goods and interests that must be internationally protected by and for the international community, delicts "only" violate the rights of a single state. In the ILC's wording,[4] an international crime consists in "*an internationally wrongful act which results from the breach by a state of an international obligation so essential for the protection of fundamental interests of the international community that its breach is recognized as a crime by that community as a whole*" (Article 19.2). Among these fundamental interests, the maintenance of international peace and security, the self-determination of peoples and global environmental protection come at play, alongside for example with the prohibition of genocide and of slavery. With regard to environmental issues only, the ILC listed among the crimes against the international community also "*a serious breach of an international obligation of essential importance for the safeguarding and preservation of the human environment, such as those prohibiting massive pollution of the atmosphere or of the seas*" (Article 19.3, letter d). However, decidedly more important in this sense is what the International Court of Justice declared in 1997. In the *Gabcikovo-Nagymaros* case, The Hague's judges explicitly pointed out that the protection of the environment is an *erga omnes* obligation

[3] It should be remembered that *erga omnes* obligations do not arise only where there is a need to face ongoing phenomena, since in reality they can also consist in negative obligations, for example with regard to the prohibition of genocide.

[4] See the Draft Articles on State Responsibility, adopted by the International Law Commission on first Reading, January 1997. https://legal.un.org/ilc/texts/instruments/english/commentaries/9_6_1996.pdf. Accessed 13 May 2019.

because of the essential importance that environmental issues have for humankind's well-being.[5]

In such a context, the safeguard of a tolerable world climate system to be attained through the achievement of the 2°C Paris Agreement objective can be easily considered as an essential global interest. The functioning of the international community and the geopolitical and socio-economic stability of states also depend on it. Indeed, a failure in mitigating global warming and, thus, a failure to achieve the objectives set out in Paris can potentially provoke a radical and historical reorganization of the international order as it presently exists. The modification of the global climate system toward excessively high temperatures, through the alteration of climatic and ecosystem balances that have endured for millennia, can lead to the outbreak of huge international conflicts, vast migratory events and widespread human suffering, which could change the international system as we know it today.

Climate change is already affecting heavily some regions of the planet; it will also do so more and more over the next decades. One crucial aim of the international community should then consist in preventing future irreparable and irreversible harm: to the environment, to the economy, and to the people(s). The precautionary principle becomes consequently relevant in this reasoning.[6] Working at the very bases of international environmental law, the precautionary principle requires states to undertake actions to prevent serious environmental harm even in the absence of full scientific certainty. At the universal level, the most widely accepted formulation of the principle is in the 1992 Rio Declaration on Environment and Development, which establishes that "*where there are threats of serious or irreversible damage, lack of full scientific certainty shall not be used as a reason for postponing cost effective measures to prevent environmental degradation*".[7] A very similar formulation is taken up in 1994 in the UNFCCC with regard only to climate change, where "*parties should*

[5] International Court of Justice (1997). *Gabcikovo-Nagymaros case (Hungary v. Slovakia)*, 25 September 1997, I.C.J. Reports 1997.

[6] See Poon on the precautionary principle applying to prevent irreparable environmental harms. Poon (2018). Drawing Upon International Refugee Law. The Precautionary Approach to Protecting Climate Change-Displaced Persons. In Behrman & Kent (Eds.) (2018). *'Climate Refugees' Beyond the Legal Impasse?* New York: Routledge, p. 157.

[7] UN Declaration on Environment and Development (Rio Declaration), adopted 12 August 1992, Principle 15.

take precautionary measures to anticipate, prevent or minimize the causes of climate change and mitigate its adverse effects. Where there are threats of serious or irreversible damage, lack of full scientific certainty should not be used as a reason for postponing such measures [...]".[8]

Under this perspective and disposition, UNFCCC's state parties are obliged to act when there is a threat of serious or irreversible damage. A relatively high threshold is consequently to be met, since we may define an irreversible damage as something that "is practically irreversible in the sense that it cannot be undone in the course of several human generations",[9] and that has a real detrimental effect on the quality of the environment. When we consider anthropogenic climate change in general, with all its associated effects—both from an environmental but, more importantly, from a socio-economic point of view—we can immediately appreciate that these damages are and will increasingly be both serious and irreversible for several human generations. So, given that global warming also impacts on the entire planet and therefore on every state, carrying also the risk to upset the functioning of the whole international community, both the fight against and the management of climate change through mitigation and adaptation policies are *erga omnes* obligations, because they aim to protect a fundamental and essential interest of the international community.

The weakness of the theorization of *erga omnes* obligations lies in the fact that reparations for damages caused by their infringement are not easily available at all for the damaged states, both "directly" and "indirectly". Directly means through pleas raised in front of international jurisdictional bodies that can coercively impose actions on the responsible subjects, whereas indirectly means through political, economic, or diplomatic reactions that can be lawfully put in place by the injured subjects "against" the state that is in violation of the law. This is even more so when it comes to planetary environmental protection, since it is difficult to clearly circumscribe the damage that occurs in one specific "victim" state, since these affect generally and indiscriminately every state in the community. Although a phenomenon like deforestation affects, generally, the well-being of every state, it is virtually impossible that one single, specific state incurs in damages that are primarily attributable to it, and

[8] UNFCCC, Article 3.

[9] Poon (2018) (op. cit.), pp. 158 and 159.

this decidedly weakens the "availability" of the respect of *erga omnes* obligations in relation to global environmental degradation. As a matter of example with regard to global warming, the reactions adopted by the collectivity of states to the withdrawal of the United States from the Paris Agreement indicate that this "availability" is quite far from grasp (particularly in reaction to the decisions of the most powerful state of the whole international order). These reactions consisted in fact so far only in a series of condemnations, criticisms, and complaints at a rhetorical level, and not in "sanctioning" reactions or retaliations against the United States.

However, the strength of *erga omnes* obligations does not lie there. What is relevant in this context is that the fight against global warming and its management have become by now an (almost) unanimously shared goal. At a purely legal level, as the UNFCCC and its implementation agreements underline, climate change is an anthropogenic phenomenon, thereby implying that a human agency exists. States must act to mitigate global warming, and given its inertial and (partially) ineluctable nature, they also have to adapt to it. If this is the case, the slow but definite appearance of a new custom that imposes certain "climatic behaviors" onto states, which include, at a *minimum*, the obligatory and progressive reduction of their greenhouse gas emissions,[10] and the establishment of (inter)national adaptation policies and strategies, can be ascertained.

The behavior adopted by the vast majority (if not the unanimity) of states in recent years—with 2015 marking the "climax" of the historical-juridical evolution under consideration here—allows in fact to appreciate that both an *usus* and an *opinio jure ac necessitatis* with regard to the obligation to mitigate and adapt to climate change are today arising. A custom appears under international law when a widespread repetition over time by states of similar international conducts (state practice or

[10]The Paris Agreement establishes that every State Party shall communicate the UNFCCC's Secretariat its greenhouse gas emission reduction five-years plans (the so-called Nationally Determined Contributions, NDC). It furthermore requests that every subsequent NDC must be more ambitious than the preceding one, generating thereby a ratchet-Up mechanism). See in particular Articles 4.2 e 4.3: "Each Party shall prepare, communicate and maintain successive nationally determined contributions that it intends to achieve. Parties shall pursue domestic mitigation measures, with the aim of achieving the objectives of such contributions"; "Each Party's successive nationally determined contribution will represent a progression beyond the Party's then current nationally determined contribution and reflect its highest possible ambition, reflecting its common but differentiated responsibilities and respective capabilities, in the light of different national circumstances".

usus), occurs out of a sense of obligation (*opinio juris*). While it is important not to merge treaty and customary law, it is nevertheless possible to assume that the existing generalized support for the international climate change regime—independently from its current effectiveness—suggests that a new customary rule (the purpose of which is to protect an essential value to the international community, thus imposing an *erga omnes* obligation) is gradually emerging within the international legal order. Also Mayer resonates similarly, when he argues that "if sovereign equality precludes a state from causing harms affecting a small part of the territory of another state, it does *a fortiori* prohibit the conduct of a state which interferes with multiple planetary systems in ways that not only affect the prosperity of many states and the very physical existence of some, but also possibly our survival as a civilization, if not as a species".[11]

If this hypothesis is valid, its scope clearly goes far beyond the particular phenomenon of "climate refugees". Limiting ourselves just to it, however, the existence of an international custom that imposes an *erga omnes* obligation to manage global warming, can be used as a further legal argument to maintain that the category of forced climate migrants should be provided with international protection in third countries.

In this regard, the embryonic evolution that can be observed over the last years in relation to the topic of climate-related migration within the UNFCCC is particularly interesting. We shall look at it thoroughly in Chapter 10, but starting from the 2010 COP 16 in Cancun the UNFCCC system has started to introduce non-binding provisions aimed at ensuring a better understanding of the phenomenon of "climate refugees" for the purposes of assuring its proper future management. Under the Cancun Adaptation Framework, states were encouraged to take "*measures to enhance understanding, coordination and cooperation with respect to climate change induced displacement, migration and planned relocation, where appropriate, the national, regional and international levels*". It is probably in reaction to this recommendation as well that, in the years immediately following Cancun, some governments gave life to the Nansen Initiative and its follow-up, the Platform on Disaster

[11] Mayer (2017). Climate Change, Migration and the Law of State Responsibility. In Mayer & Crépeau (Eds.), *Research Handbook on Climate Change, Migration and the Law*. Cheltenham: Elgar, p. 246.

Displacement,[12] both of which are intergovernmental initiatives that have had great success in clarifying the nature of the phenomenon, above all from the viewpoint of national practices that are in place today to assist people in the event of natural disasters. More recently, the legacy of COP 16 was carried forward in Paris in 2015, when, in the Preliminary Decision to the Paris Agreement, parties decided to include the subject within the "loss and damage" section, providing for the creation of a task force with the objective of studying and minimizing the phenomenon of migrations related to global warming.[13]

Such dispositions can be viewed as the embryo of a "climate refugees law" and can, therefore, also be understood as the fruit of the progressive emergence of an international responsibility by virtue of which, on the wave of the ever-greater awareness regarding the human origin of climate change, a shared management system of the phenomenon of "climate refugees" should be put in place. Building on this, Atapattu points out for instance that "the obligations under the UNFCCC legal regime are *erga omnes* and this, coupled with human rights law, provides us with a good framework to protect the rights of climate displaced persons".[14] The objective of the next chapter is then precisely to understand how human rights law fits into our discussion so far. If we link the obligations that states have in respecting and protecting non-derogable human rights with the ones they bear concerning climate change mitigation and adaptation, there is a good chance of making a convincing argument for the existence of a duty to protect forced climate migrants.

[12] For more information on the activity of the *Platform on Disaster Displacement*, see the website https://disasterdisplacement.org/.

[13] For further information on the evolutions occurred under the UNFCCC's auspices on the topic of climate-migration, please refer to Chapter 10.

[14] Atapattu (2018). A New Category of Refugees? "Climate Refugees" and Gaping Hole in International Law. In Behrman & Kent (Eds.), *"Climate Refugees" Beyond the Legal Impasse?* New York: Routledge, p. 48.

CHAPTER 8

Climate Change, Environmental Degradation, and Fundamental Human Rights

"There is [not] a complete *lacuna* in international law. As with other migrants, those displaced by environmental factors enjoy all of the human rights applicable [...]. The Universal Declaration of Human Rights, the International Covenant on Civil and Political Rights and the International Covenant on Economic, Social and Cultural Rights define the basic rights of all persons".[1] Hence, as Martin underlines, when we study the potential applicability of protection frameworks to the category of forced climate migrants, we are fortunately not in a complete *vacuum*.

Thus, in the reasoning carried out so far, there is a further aspect that is linked to the concept of international responsibility, namely that concerning the obligations that states have in protecting fundamental human rights. Human rights law is in fact "the only existing internationally agreed expression of the minimum conditions that everyone should enjoy if they are to live with dignity as human beings"[2] and it therefore establishes a common and universal standard that should be applied

[1] Martin (2017). Toward an Extension of Complementary Protection? In Mayer & Crépeau (Eds.), *Research Handbook on Climate Change, Migration and the Law*. Cheltenham: Elgar, p. 449.

[2] Van der Vliet (2018). "Climate Refugees", a Legal Mapping Exercise. In Behrman & Kent (Eds.), *"Climate Refugees" Beyond the Legal Impasse?* New York: Routledge, p. 26, referring to Morel (2014). *The Right Not to Be Displaced in International Law*. Cambridge: Intersentia.

G. Sciaccaluga, *International Law and the Protection of "Climate Refugees"*,
https://doi.org/10.1007/978-3-030-52402-9_8

everywhere and to everyone. In this sense, van der Vliet maintains that "as the human rights framework applies everywhere for everyone, it applies to all climate refugees. It can thus be beneficial to frame climate-induced migration as a human rights violation, as this may provide legally binding obligations for states and legally enforceable entitlements and rights for individuals. Although not designed to deal specifically with climate refugees, existing human rights law deals with a range of situations that respond to their needs and rights".[3]

As we pointed out before, global warming—also in its legal understanding—is considered as being caused by human activities and it is hence possible to identify to a certain extent human agents that are responsible for its creation and effects. And it is nowadays acknowledged that climate change can affect negatively the correct exercise of both collective and individual human rights.[4] In this chapter, we are going to focus on those rights whose exercise can be compromised in the context of migrations that are forced by climate change-related environmental degradation. The intent is to show how global warming, when impacting on certain fundamental rights that are universally recognized by the international community, can determine—obviously in a multifaceted context—forced migrations, which in turn to generate a need for international protection. In a world with a changing climate, not only individual rights come at stake. Stakeholders and policy-makers should also not underestimate those collective rights whose enjoyment may be jeopardized, above all when we consider insular micro-states at risk of

[3] Ibid., p. 21.

[4] The link between fundamental rights and climate change has been repeatedly acknowledged within the international community. There is no question that the negative effects of climate change can affect the individuals' enjoyment of their most basic rights. In this regard, for example, the UN Human Rights Council has identified as possible causes of deprivation of fundamental rights the following environmental phenomena: coastal erosion; sea tides; floods; drought; sea-level rise; storms, hurricanes, cyclones. Through the destruction of the political, economic, institutional structures of a community, rights such as the right to life, to health, housing, work, to culture, access to, livelihoods, and even to self-determination are likely to be put at risk. See UN Human Rights Council (2009). Report of the Office of the United Nations High Commissioner for Human Rights on the Relationship Between Climate Change and Human Rights. https://documents-dds-ny.un.org/doc/UNDOC/GEN/G09/103/44/PDF/G0910344.pdf?OpenElement. Accessed 12 January 2020.

submersion.[5] For these subjects new legal challenges arise, which even go as far as to integrate their possible extinction as international law subjects, given their foreseeable and not unlikely "deterritorialization" which will occur in the future due to sea-level rise.

In the present chapter, we focus however on the link between climate change and the exercise of basic individual rights, such as the right to life, to health, and to access means of subsistence. In addition, due to the nature of our topic, we also have to consider the role of the right to live in a healthy or decent environment. For an analytical purpose, we in fact begin by considering this last human right first, to proceed later on with the other, more canonical and less discussed rights.

The Environmental Dimension of Fundamental Rights

When we consider here the human right to live in a decent environment, the intent is not so much to show how "forced climate migrants" could enjoy international protection rights by virtue of this specific right, but rather to conceive a healthy environment as an element that is inevitably necessary for the effective enjoyment of other universally recognized fundamental rights. Therefore, the focus is not on the right of the environment to be healthy or decent *per se*. We rather consider the right to a clean environment as that set of rules, both of substantive and procedural nature, which oblige states to protect a certain quality of the environment so as to guarantee that basic human rights can effectively be exercised and enjoyed within their territories. In this perspective, we are not talking of a fundamental right to a clean environment, but rather of an environmental dimension of fundamental rights.

[5] See for instance Poon on this subject: "The Pacific Islands are a group of 11 Pacific Island countries, such as Fiji, Marshall Islands and Papua New Guinea, which are most vulnerable to climate change displacement. The Pacific Islands face a serious threat to their social and political stability due to constant rise in sea levels as a result of impacts of climate change [...]. Those who live on Pacific Islands will be forced to continue to relocate internally, if possible, or worse, across borders as a result of climate change displacement. In fact, based on 2009 estimates, it is projected that 1.7 million people will be affected by climate change displacement by 2050 from the Pacific Islands alone". In Poon (2018). Drawing Upon International Refugee Law. The Precautionary Approach to Protecting Climate Change-Displaced Persons. In Behrman & Kent (Eds.), *"Climate refugees" Beyond the Legal Impasse?* New York: Routledge, p. 165.

The concept of a right to a clean, healthy, decent, or satisfactory environment has progressively arisen with the evolution and re-elaboration of the "classic" human rights that are embodied within the 1948 Universal Declaration of Human Rights. More precisely, the idea of a right to live in a clean environment comes with the "third generation human rights".[6] Its origin can be traced to the first principle of the 1972 Stockholm Declaration, which states that "*Man has the fundamental right to freedom, equality and adequate conditions of life, in an environment of a quality that permits a life of dignity and well-being, and he bears a solemn responsibility to protect and improve the environment for present and future generations*".

This principle clearly recognizes the existence of a right to live in a decent environment, and represents the starting point from which this subject has evolved in the subsequent years and decades. By referring to a solemn responsibility to protect future generations, it also recalls, more generally, an *erga omnes* obligation to protect the environment. Moreover, by calling for interstate rights and duties, it imposes a solemn responsibility that humankind has in protecting the environment, understood as a common and shared good. The Stockholm Declaration is to a certain extent quite radical in its nature and language, because it adopts a clear universalist approach in which explicit reference is made to issues of intergenerational responsibility.[7]

Despite (or because) of its language, the nature of the Declaration has probably been "betrayed" by its direct heirs: the following major global conferences dedicated to environmental matters have in fact gradually "softened" their approach, losing part of the more purely "ecological" momentum that appears in the 1972 Declaration. The outcomes of the 1992 Rio de Janeiro Conference for instance clearly show that environmental issues had to be integrated at that time with economic and, let us say, "materialistic" interests. The concerns raised particularly, but not exclusively, by developing countries with regard to their economic growth needs changed the language and the political outcomes of 1992, to the

[6] For a detailed analysis, see Munari & Schiano di Pepe (2012). *Tutela Transnazionale dell'Ambiente*. Bologna: il Mulino, Chapter 3.

[7] See in this sense: Francioni (2014). Comments on the Preamble and on Principle 1. In Viñuales (Ed.), *The Rio Declaration on Environment and Development: A Commentary*. Oxford: Oxford University Press. See also Picone (2013). *Comunità Internazionale ed Obblighi Erga Omnes*, Naples: Jovene Editore, pp. 18–19.

point of interpreting environmental issues and the management of natural resources in a decidedly more "developmental" perspective than it was the case in 1972.[8]

Despite this involution, international law has witnessed a growing affirmation of the role of a fundamental right to a healthy environment. New soft and hard law instruments and jurisprudential evolutions of preexisting dispositions can be found all over the world.

With respect to soft law, a particular role is played for instance by the 1989 Declaration on the Environment of The Hague, which established a close relationship between a healthy environment and the right to life.[9] Similarly, a 1990 UN General Assembly's resolution stated that "*all individuals are entitled to live in an environment adequate for their health and well-being*".[10] In 1994, the UN Subcommittee on the prevention of discrimination and the protection of minorities adopted a Draft Declaration of Principles on Human Rights and the Environment that is particularly interesting even though it received little support by states.[11] And this is above all because of its radical content and language. In fact, the Draft follows an "ecologically strong" approach, on the same model of the 1972 Stockholm Declaration. It recognizes the existence of an *intergenerational* right to an ecologically healthy environment, which is moreover *interdependent* and *indivisible* with respect to other human rights.[12] It recognizes an individual right, also endowed

[8] See in this regard: Nespor (2009). *Il Governo dell'Ambiente.* Milan: Garzanti. See also Zaccal & Lugen (2016). Common But Differentiated Responsibilities Against the Realities of Climate Change. In Papaux & Zurbuchen (Eds.), *Philosophy, Law and Environmental Crisis.* Stuttgart: Nomos.

[9] The Hague Declaration on the Environment, adopted on 11 March 1989, 28 I.L.M. 1308.

[10] UNGA Resolution (1990). Need to Ensure a Healthy Environment for the Well-Being of Individuals, UNGA Res 45/94, A/RES/45/94.

[11] Draft Principles on Human Rights and the Environment, adopted 16 May 1994, E/CN.4/Sub.2/1994/9, Annex I.

[12] Ibid., Principles 2 and 4. On the subject of intergenerational equity, the positions defended by the United States in occasion of the 1893 *Bering Sea Fur Levels Arbitration* case can be interesting: "The earth was designed as the permanent abode of man through ceaseless generations. Each generation, as it appears upon the scene, is entitled only to use the fair inheritance. It is against the law of nature that any waste should be committed to the disadvantage of succeeding tenants [...]. That one generation may not only consume or destroy the annual increase of product of the earth, but the stock also, thus leaving an inadequate provision for the multitude of successors which it brings to life, is a notion

with extraterritorial scope, to freedom from pollution and activities that cause environmental degradation, and it furthermore envisages states obligations to protect and safeguard air, soil, water, oceans, and biodiversity.[13] Unsurprisingly and probably precisely because of this radical approach which excessively focuses on environmental responsibility, the Draft declaration has never found general support from the international community.

At the same time, while some regional human rights courts, such as the Court of Strasbourg,[14] began to interpret preexisting dispositions in an evolutionary and "green" way, establishing the existence of an implicit but clear right to a healthy environment, human rights dedicated bodies have also expressed themselves explicitly on the subject matter.[15]

This is the case, for example, of the African Charter on Human and Peoples' Rights of 1981, whose Articles 16 and 24 establish the right of peoples to enjoy the "*best attainable standard of health*", and a generally satisfactory environment, favorable to their development. It is worth highlighting that these articles are not formulated in an individual perspective, but rather in a collective one, since they refer to the "right of peoples". The infamous 2001 *Ogoniland case* in Nigeria still represents the most important judgment pronounced so far on this question. The Human Rights and Peoples' Rights Commission of the Charter convicted the Nigerian government for violations of the aforementioned provisions for having allowed on its own territory—and in the area where

so repugnant to reason as scarcely to need formal refutation". Cit. in Spier (2016). Intergenerational Equity: An Aspiration or an Effective Weapon? In Papaux & Zurbuchen (Eds.), *Philosophy, Law and Environmental Crisis*. Stuttgart: Nomos, p. 69.

[13] Draft Principles on Human Rights and the Environment, adopted 16 May 1994, E/CN.4/Sub.2/1994/9, Annex I., Principles 5 e 6.

[14] For further information and considerations, referring in particular to the European context, see Pavoni (2014). *Interesse pubblico e diritti individuali nella giurisprudenza della Corte europea dei diritti umani*. Naples: Editoriale Scientifica.

[15] Cf. Knox (2015). UN Human Rights Council's Report of the Independent Expert on the Issue of Human Rights Obligations Relating to the Enjoyment of a Safe, Clean, Healthy, and Sustainable Environment. Compilation of Good Practices (A/HRC/28/61). See also UN OHCHR (2014). Mapping Human Rights Obligations Relating to the Enjoyment of a Safe, Clean, Healthy, and Sustainable Environment: Focus Report on Human Rights and Climate Change. Mapping Report. UNHCR. https://www.ohchr.org/EN/Issues/Environment/SREnvironment/Pages/MappingReport.aspx. Accessed 22 January 2019.

the Ogoni ethnic minority used to live (and where they still live)—hydrocarbon extraction activities in the most blatant disregard of any legislation protecting the environment and the rights of peoples and individuals. The Commission concluded that the parties to the Charter are responsible for preventing pollution and ecological degradation, for guaranteeing the conservation of an ecologically sustainable environment and for ensuring an economic development that is sustainable in the short and medium run.[16]

Moving to the Americas, Article 11 of the 1998 San Salvador Protocol of the American Convention on Human Rights is also clear, establishing an individual right to live in a healthy environment. Holding a similar position, we can also find Article 38 of the Arab Charter on Human Rights, according to which "*every person has the right to an adequate standard of living for himself and his family, which ensures their well-being and a decent life, including food, clothing, housing, services and the right to a healthy environment*".[17] Similarly, the ASEAN Declaration on Human Rights of 2012 establishes that "*every person has the right to an adequate standard of living for himself or herself and his or her family including the right to a safe, clean and sustainable environment*".[18] Finally, Article 37 of the Charter of Fundamental Rights of the European Union is also relevant. While not recognizing an explicit individual right to a clean environment, it sets out, in the EU's human rights treaty, that a high level of environmental protection, together with its improvement and the pursuit of sustainable development, must be integrated and taken into account in the implementation of any European Union's policy. Article 37 has so far never been applied by the European Court of Justice, surely because of its indirect wording. In fact, it seems that the EU legislator's objective

[16] See *Social and Economic Rights Action Centre (SERAC) and Another v Nigeria* (2001) AHRLR 60 (ACHPR 2001), point 52. See also point 58: "The [African] Commission notes that in the present case, despite its obligation to protect persons against interferences in the enjoyment of their rights, the Government of Nigeria facilitated the destruction of Ogoniland. Contrary to its Charter obligations and despite such internationally established principles, the Nigerian Government has given the green light to private actors and the oil companies in particular, to devastatingly affect the well-being of the Ogonis".

[17] Arab Charter on Human Rights, adopted 15 September 1994.

[18] ASEAN Declaration on Human Rights, adopted 9 November 2012, para. 28.

was to further legitimize the Union's environmental integration principle within a human rights instrument, rather than establishing a direct, straightforward individual right to live in a healthy environment.[19]

In a complementary way, the individual right to a healthy environment has been indirectly guaranteed through evolutionary interpretations of preexisting human rights provisions, whose exercise is precluded when some environmental contexts are seriously inadequate. In this perspective, the action of the European Court of Human Rights is exemplary. The Court relies on Articles 2 and 8 of the European Convention on Human Rights (ECHR)—that, respectively, protect the right to life and to family and private life—and "greens" them. Over the years, the Court of Strasbourg has "greened" these articles and interpreted them in relation to their environmental dimension. With regard to the right to life for instance, the Court expressed itself clearly in two landmark cases: *Öneryildiz v. Turkey*[20] and *Budayeva v. Russia*.[21] In both, the two defendant states were convicted because of violations of ECHR's Article 2. While aware of the risks associated with particular environmental situations under their jurisdictions,[22] the two states did not take all the necessary measures, complying with their positive obligations under the ECHR, to avoid casualties.[23] In the judges' eyes, these omissions have indirectly violated the right to life, which the parties are required to protect from any activity, both public and private, that is under their jurisdiction.

[19] Charter of Fundamental Rights of the European Union, Article. 37. Article 11 of the Treaty on the Functioning of the European Union is relevant in this subject matter. It establishes that environmental protection objectives and environmental concerns must be integrated into any other Union's policy.

[20] European Court of Human Rights, *Öneryildiz v. Turkey*, Application no. 48939/99, 30 November 2004.

[21] European Court of Human Rights, *Budayeva e al. v. Russia*, Applications no. 15339/02, 21166/02, 20058/02, 11673/02 e 15343/02, 20 March 2008.

[22] Respectively: (i) having allowed that some households lived on an underground garbage dump that eventually exploded due to excessive gas pressure, and (ii) having failed in taking measures to protect a population subject to the risk of a widely predicted landslide.

[23] These are normative obligations, also of administrative nature, which imply both the adoption of substantial rules and standards aimed at limiting the risks of loss of human lives and the introduction of control and verification procedures on compliance, by the interested parties.

However, it is Article 8 of the Convention, in protection of family and private life, which is more often called into play in relation to environmental quality issues. This happens given the assumption that a normal private or family life cannot be exercised in decidedly unfavorable environmental conditions, such as serious acoustic, atmospheric, soil, or water pollution. The Court of Strasbourg has repeatedly ruled that ECHR's Article 8 also applies to protect individuals from situations of "*serious environmental risk*" and, more precisely, from a "*harmful effect on a person's private or family sphere and not simply the general deterioration of the environment*".[24] This happened for example in relation to *Guerra v. Italy*,[25] where Italy was condemned for not having taken measures allowing the local population to access information on the dangerous activities of a near petrochemical plant. Similarly, in *Taskin v. Turkey*, Turkey was convicted for violations of Article 8 following the omission of protective measures requested by the applicant aimed at interrupting mining activities that were harmful to the environment in which he regularly lived. There are many other cases where state parties were condemned for failing to take positive action aimed at preventing the occurrence of serious environmental risks capable of affecting health, life, or private property.[26]

The so-called "proceduralization of environmental rights", or to what is sometimes labeled as "environmental democracy", deserves special attention. In fact, the Strasbourg Court has often condemned states for Article 8 violations when they fail to comply with the procedural requirements that are necessary to ensure that certain economic activities (mostly of an industrial nature) are respectful of the environment and of the private life of those living in the environment where these activities are carried out. In *Giacomelli v. Italy* for instance, the Court held that Article 8 can be applied for ascertaining whether the decision-making process

[24] See in this sense the European Court of Human Rights, *Kyrtatos v. Greece*, Application no. 41666/98 (point 52), 22 Amy 2003; *Fadeyeva v. Russia*, Application no. 55723/00, 30 November 2005 (point 68); *Di Sarno* et al. *v. Italia*, Application no. 30765/08, 10 January 2012 (point 80).

[25] European Court of Human Rights, *Guerra v. Italia*, Application no. 14967/89, 19 February 1998.

[26] European Court of Human Rights, *Fadeyeva v. Russia*; *Hatton v. United Kingdom*, Application no. 36022/97, 10 July 2003; *Moreno Gomez v. Spain*, Application no. 4143/02; *Faegerskioeld v. Sweden*, Application no. 37664/04, 26 February 2008.

with which the installation of a waste disposal plant had adequately addressed the private interests of local populations or individuals.[27]

The basic rationale of such a system consists in the idea that in presence of activities with strong environmental impact, an "acceptable" environmental quality can be guaranteed only when certain procedural requirements are met. Among these, access to information, the right to participate in decision-making processes, and, more generally, the right to access justice are the most important ones. The aforementioned Ogoni case in Nigeria also referred to this point, arguing that "*government compliance with the spirit [...] of article 24 [which regulates the right to a healthy environment within the environment to the African Charter] must also include ordering or at least permitting independent scientific monitoring of threatened environments, requiring and publicizing environmental and social impact studies prior to any major industrial development [...]*".[28] In the Americas, the Inter-American Court of Human Rights condemned Belize for having allocated, in violation of the right to property of some indigenous populations, concessions for the exploitation of natural resources, and for failing to guarantee adequate judicial protection to the indigenous community involved (see *Maya Indigenous Communities v Belize.*[29])

Moreover, there is an international treaty (mainly European, but actually open to any state) specifically dedicated to "environmental democracy": the 1998 Aarhus Convention on access to information, citizen participation and access to justice in environmental matters.[30] Access to information, citizen participation in decision-making and access to justice are, indeed, the pillars of the "procedural" right to a healthy environment. A series of procedural requirements is imposed on state authorities to ensure that each activity—both public and private—carried out in their

[27] European Court of Human Rights, *Giacomelli v. Italia*, Application no. 59909/00, 19 October 2006.

[28] *Social and Economic Rights Action Centre (SERAC) and Another v. Nigeria* (2001) AHRLR 60 (ACHPR 2001), para. 53.

[29] Inter-American Court of Human Rights, *Maya indigenous community of the Toledo District v. Belize*, case 12.053, report no. 40/04, para. 193–197.

[30] Convention on Access to Information, Public Participation in Decision-Making, and Access to Justice in Environmental Matters, adopted 25 June 1998, entered into force 30 October.

territory does not produce harmful effects on the quality of the environment. In this perspective, the violation of the individual right to a clean environment is not directly related to the environmental degradation in itself, but to the "failure to comply with procedural rules to allow verification of the circumstance if this change [or degradation] was necessary and proportionate to the objective that was intended to be achieved with this activity, and consistent with a balance of opposing interests".[31]

If, globally speaking, it is clear that a purely procedural dimension of the right to live in a healthy environment exists and applies, it is also true that violations of the fundamental right to live in a decent environment not only occur in presence of formal and procedural abuses. A growing number of national constitutions refer to a substantive right to the environment, not only to a procedural or formal one. The procedural dimension of the right to a clean environment seems, indeed, to be just a particular facet of the fundamental right to a healthy environment, that applies in fact, both internationally and nationally, in more and in different ways. Indeed, 177 countries have recognized a right to live in a healthy environment, be it at constitutional level, through to ordinary law, jurisprudential interventions, or through the ratification of international treaties. More precisely, 110 states present a constitutional guarantee, and 120 have it through international law instruments.[32] In the light of all this, this author totally agrees with the idea that "were the Universal Declaration [on Human Rights] adopted today, it would include a right to [a healthy] environment, recognized in so many national constitutions and regional agreements".[33]

If we also consider what the International Court of Justice stated back in 1997, namely that environmental protection represents an *erga*

[31] Munari & Schiano di Pepe (2012). *Tutela Transnazionale dell'Ambiente*. Bologna. il Mulino, p. 128 (translation from Italian by this author). Please also refer to Chapters 3.2.4 and 3.2.5 of the aforementioned work for a more complete and in-depth examination of the procedural dimension of the right to a healthy environment, both at the international and European Union level.

[32] Cf. Francioni & Quirico (2016). Untying the Gordian Knot. In Quirico & Boumghar (Eds.), *Climate Change and Human Rights*. New York: Routledge, p. 147. See also UNEP et al. (2014). Regional Consultation on the Relationship Between Human Rights Obligations and Environmental Protection, with a Focus on Constitutional Environmental Rights. See also Boyd (2012). *The Environmental Rights Revolution*. Vancouver: UBC Press.

[33] Cf. Knox (2015) (op. cit.).

omnes obligation since environmental issues are of essential interest to humanity,[34] we can conclude that the right to a clean environment (even if clearly declined in different ways) is insistently becoming part of international general law,[35] and that its current incomplete recognition is due to its historical freshness and current fragmentation.

In any case, the fundamental right to the environment (whatever its effective implementation) does not concretize as something intended to protect the environment *per se*, but rather as a tool aimed at ensuring other fundamentals rights of individuals—life, health, family life, property, and so on. A damage to the environment does not constitute in itself a violation of fundamental rights: we in fact normally speak of a right to a clean environment, not of a right of the environment *per se*. More accurately, in reality, we should talk of the "acceptable enjoyment of rights within a given environment" or of the "environmental dimension of fundamental rights". When public authorities allow decisive degradation of the environment in which these rights are exercised, human right infringements or violations can occur. In this sense—and as obvious, given the liberal origins of the legal world we live in—the right to the environment is clearly understood in an anthropocentric and, very often, individualistic way, putting human life at the center, and not, therefore, the environment itself. In all this, nonetheless, a certain level of environmental protection is indispensable for the effective exercise of universally recognized individual rights.

Having unpacked the link between environmental quality and human rights protection, we need to clarify how this fits with climate change and, eventually, with forced climate migrants.

Fundamental Human Rights and Forced Climate Migrants

It is generally acknowledged now that the effects of global warming can negatively affect the exercise of certain basic human rights, such as the right to family life and property for instance, but above all the rights to life

[34] International Court of Justice, *Gabcikovo-Nagymaros case* (cit.), p. 7, para. 41. See also: International Court of Justice, *Advisory opinion on the Legality of the Threat or Use of Nuclear Weapons*, I.C.J. Reports 1996, p. 226.

[35] See in this regard Boyd (2012). The Constitutional Right to a Healthy Environment. *Environment: Science and Policy for Sustainable Development*, 4, p. 3 ff.

and health. These fundamental rights, whose origin can be traced down in the 1948 Universal Declaration of Human Rights, are enshrined, at a universal treaty level, in the two 1966 Covenants on civil and political rights, and on economic, social, and cultural rights.[36] They are moreover specifically protected by regional human rights treaties, such as the European Convention on Human Rights, the American Convention on Human Rights, and the African Charter for Human and Peoples' rights. Therefore, their nature as fundamental rights that are part of general international law is not to be called into question.

When a person is compelled to leave his or her own country because of climate change and associated unbearable environmental conditions, there is no doubt that a significant level of human suffering is occurring, susceptible (at least in theory) of integrating infringements of universally recognized human rights. Serious environmental degradation, of climatic, geophysical, or meteorological origin, can in fact condemn some individuals—or even entire communities—to live in unacceptable socio-economic conditions. Just consider, for example, the difficulties in accessing basic livelihoods in areas repeatedly subject to severe drought, desertification, rising seas, or salinization of aquifers.

If the enjoyment of the basic substantive rights is of primary importance, we should also not underestimate those procedural rights (such as the right to participate in decisions concerning the displacement of homes or families due to emergencies), which are for example clearly recognized in the 1998 Guiding Principles on Internal Displacement.[37] Although it remains debatable that these procedural rights enjoy widespread international reach, with protection mechanisms capable of ensuring compliance in the majority of states, it remains equally clear that, where these rights are or should be guaranteed, their respect could be at risk in cases of numerically exorbitant forced migrations.

Alongside these concerns, global warming can also have negative effects on the enjoyment of the collective rights of some particularly vulnerable populations. The right to culture and the right to nationality could, for instance, face severe challenges for the populations of those

[36] International Covenant on Economic, Social, and Cultural Rights, adopted on 16 December 1966, entered into force 3 January 1976.

[37] UN High Commissioner for Refugees (UNHCR) (1998). Guiding Principles on Internal Displacement. ADM 1.1.PRL. 12.1, PR00/98/109.

micro-state islands threatened by submersion. Therefore, there are essentially two dimensions of rights that must be considered when thinking of the phenomenon of "climate refugees", individual, and collective ones.[38]

Historically speaking, the association of anthropogenic climate change with the enjoyment of human rights is recent. Originally, in fact, global warming was mainly considered a purely environmental topic, with the consequence that studies and actions dealing with it were for many years largely conducted by actors specialized in environmental matters. Thus, although climate change has been part of the international political debate for several decades now, it is only since the late 2000s that human rights dedicated international institutions began to take an interest in it.[39] For instance, the UN Human Rights Council expressed its opinion in this regard only in 2008, and it acknowledged that climate change represents an immediate and large-scale threat to individuals and communities around the world, with implications for the full enjoyment of human rights, also underlining the particular vulnerability of certain geographical areas.[40] In turn, the Office of the High Commissioner for Human Rights (OHCHR) also dedicated some time to the topic, concluding that the enjoyment of different rights, be they civil and political, or economic and cultural, can be challenged by global warming.[41] Among these rights, the rights to life, food, water, health, home, and self-determination were

[38] See Atapattu (2016). *Human Rights Approaches to Climate Change: Challenges and Opportunities*. New York: Routledge, p. 68: "The rights affected include the right to life, right to property and the right to a livelihood, while those living on small island states are facing the risk of extinction as their island nations are threatened by submersion associated with sea-level rise. Other communities are facing the likelihood of losing their traditional way of life, their culture and their traditional lands".

[39] For a detailed explanation as to why this happened, see McInerney-Lankford (2013). Human Rights and Climate Change. In Gerrard & Wannier (Eds.), *Threatened Island Nations: Legal Implications of Rising Seas and a Changing Climate*. New York: Cambridge University Press, pp. 229–230: "The international bill of rights is a product of the post-World War II era and is oriented toward preventing specific moral harms and preserving a set of substantive and procedural entitlements for all. The climate change regime for its part emerged much later, as the outcome of very different movements and awareness. It is aimed at tackling harms that were not contemplated in the design of human rights frameworks, which are therefore ill equipped to tackle them".

[40] See UN Human Rights Council Resolution 7/23 (2008). Human Rights and Climate Change. https://ap.ohchr.org/documents/E/HRC/resolutions/A_HRC_RES_7_23.pdf. Accessed 15 March 2019.

[41] UN Human Rights Council (2009). Report of the Office of the United Nations High Commissioner for Human Rights on the Relationship Between Climate Change

particularly stressed. Moreover, it is also interesting to note that global warming, given its extremely complex nature, can impact on human rights not only through its direct effects, but also, paradoxically, through the actions put in place to mitigate it.[42] The production of biofuels for example can have negative effects on the rights of populations and individuals in relation to their access to food, where land grabbing activities are undertaken in disrespect of the rights of those who live in the "grabbed" areas of the planet.[43]

Regardless of this peculiar consideration, a specific analysis is necessary in relation to those fundamental rights, such as the right to life, to health, and access to livelihoods, that is potentially and actually endangered by global warming, so as to better understand the threshold beyond which climate change-driven environmental degradations may effectively force someone to leave his or her country, making him or her a forced climate migrant.

Right to Life and Climate Change

First, the right to life plays a pivotal role. According to the studies of the IPCC, climate change, especially through the intensification and the increase of extreme weather events can contribute to causing large numbers of deaths. Today, with an increase in the average global temperature compared to 1850 of only 1°C, there are already numerous cases of cyclones, heat waves, fires, droughts, and floods of unprecedented force throughout the whole planet.[44] These events obviously cause many human casualties. It remains however extremely difficult, if not completely impossible, to attribute the main responsibility for these events to climate

and Human Rights. A/HRC/10/61. https://www.refworld.org/docid/498811532.html. Accessed 7 September 2019.

[42] Although not specifically referring to human rights, the Paris Agreement' highlights this possibility as well in its Preamble: "Recognizing that Parties may be affected not only by climate change, but also by the impacts of the measures taken in response to it".

[43] In this regard, please also refer to Atapattu (2016) (op. cit.), p. 71.

[44] According to OHCHR data, between 1980 and 2000 tropical cyclones have caused the death of approximately 250,000 individuals, while between 2000 and 2004 the lives of approximately 262 million people (98% of whom reside in developing countries) were affected negatively every year from disasters in whose manifestation climate change played a significant role. See UN OHCHR (2014). Mapping Human Rights Obligations Relating to the Enjoyment of a Safe, Clean, Healthy and Sustainable Environment: Focus Report

change, that only affects the probability and the intensity of extreme weather events. The resilience of the affected communities varies considerably independently of any climate change-associated effect. Therefore, while it is impossible to consider climate change as the only, single, and main cause of the deprivation of the preliminary and mandatory right to life,[45] it is certainly possible to conclude that it plays an increasingly significant role in the integration of this "violation". Alongside this, the right to life can be infringed where there are drastic negative ecosystemic alterations that affect the capability of certain areas to provide food and water to populations through the ecosystem services we mentioned in the Chapter 1.

Considered by some as "the foundation of humankind as a member of a society",[46] the right to life must be guaranteed by states through the fulfillment of both negative and positive obligations. States are obliged to refrain from arbitrarily depriving individuals from their lives, and—more importantly with a view to climate change—to ensure that their territories offer all those social and economic conditions thanks to which the right to life can be ultimately respected.

With regard to universal international law, the right to life is protected by Article 6 of the Covenant on Civil and Political Rights (ICCPR) for instance. The Covenant establishes that the right to life is inherent to every human being and that the law must protect it. The UN Human Rights Council has declared that states, in the implementation of their positive obligations, are called to take all necessary measures to protect human life.[47] The right to life finds moreover a special discipline within the Convention for the Rights of the Child (CRC), again in Article 6. At a regional, and hence complementary level, Article 2 of the European Convention on Human Rights is relevant. In the Americas, Articles 1 and 4 of the American Convention on Human Rights play a role, by imposing

on Human Rights and Climate Change. Mapping Report. UNHCR. https://www.ohchr.org/EN/Issues/Environment/SREnvironment/Pages/MappingReport.aspx. Accessed 22 January 2019.

45 International Covenant on Economic, Social, and Cultural Rights, Article 6.

46 Tomuschat (2010). The Right to Life: Legal and Political Foundations. In Tomuschat, Lagrange, & Oeter (Eds.), *The Right to Life*. Leiden: Brill, p. 3.

47 UN Human Rights Committee (HRC) (2018). General comment no. 36: Article 6 (Right to Life), para. 62. https://www.refworld.org/docid/5e5e75e04.html. Accessed 7 May 2020.

on states the establishment of all public measures and policies aimed at ensuring that the right to life can be guaranteed.[48] The African Charter then sanctions the inviolability and irrevocability of the right to life in Article 4.[49]

Consequently, therefore, every state should, once proven the potential negative impact of climate change on the exercise of the right to life, be under the obligation to strive to mitigate global warming, as well as to promote adaptation policies with the overall objective of ensuring that the right to life is respected under its jurisdiction. This is indisputably true at a theoretical level. In practice however, international law is often devoid of instruments to enforce human rights law. It thus becomes difficult to clarify the extent to which states are actually required to fulfill their obligations, especially when entering the field of climate change, with all the international and intergenerational complexities and uncertainties that characterize it.

In this sense, the OHCHR, while highlighting the negative implications that climate change has on the enjoyment of some fundamental rights, has questioned to what extent these implications can be considered classic violations of human rights.[50] In this regard, in fact, the never ending difficulty encountered in proving the existence of a direct causal link between climate change, extreme natural events, and violations of human rights was clearly pointed out. An even higher difficulty was noted for the case of gradual environmental degradation processes, where the temporal discrepancy between the foreseen events (for example the disappearance of an island) and the damage caused at the present moment entails the impossibility to require a jurisdictional remedy at this time.

In January 2020, the UN Human Rights Committee issued a landmark decision dealing with a climate-related migration case raising human

[48] Again, limiting ourselves to some cases in which the problem of a healthy environment and the exercise of the right to life has arisen, we can refer to the rulings of the Inter-American Court of Human Rights *Yanomani Indians v Brazil,* case no. 7615, 3 March 1985; *Kawas-Fernandez v Honduras*, 3 April 2009; *Maya Indigenous Community of the Toledo District v Belize*, case no. 12.053, 12 October 2004.

[49] See the Ogoniland case (cit.). See also *Norbet Zongo v. Burkina Faso*, appl. 13/2011, March 2014.

[50] UN Human Rights Council (2009). Report of the Office of the United Nations High Commissioner for Human Rights on the Relationship Between Climate Change and Human Rights. A/HRC/10/61. https://www.refworld.org/docid/498811532.html. Accessed 7 September 2019, para. 70.

rights concerns, where it expressly highlighted that *non refoulement* obligations should apply when asylum seekers risk their life in the country of origin because of climate-induced environmental degradations.[51] The Committee was asked to express itself on the renowned "Teitiota case", which deals with an I-Kiribati asylum seeker who has failed, over the last years, to obtain protection in New Zealand from climate-related hazards that made it allegedly impossible for him and his family to decently live in Kiribati (in this specific case in the isle of Tarawa). As a result of these failures, Mr. Teitiota has been repatriated (*refoulé*) to Kiribati by New Zealand authorities, which have held that environmental degradations "*do not convert the unhappy position of the applicant and other inhabitants of Kiribati into points of law*",[52] thereby falling outside the scope of application of the 1951 Refugee Convention. In response, Mr. Teitiota appealed to the UN Committee, claiming that, by deporting him back to Kiribati, New Zealand had violated his right to life under the ICCPR because of the detrimental effects that sea-level rise is having on the country, such as, for instance, scarce water availability and difficulty in growing crops and other means of subsistence.

After thorough consideration of the case's elements and of the behavior of New Zealand's authorities throughout the years, the Committee eventually dismissed Mr. Teitiota's plea too. In the Committee's view, his claim that life in Tarawa is at high risk because of a situation of generalized violence and severe environmental degradation provoked by sea-level

[51] UN Human Rights Committee (HRC) (2020). *Ioane Teitiota v. New Zealand* (advance unedited version), CCPR/C/127/D/2728/2016, 7 January 2020. https://www.refworld.org/cases,HRC,5e26f7134.html. Accessed 7 June 2020. On which see, *ex multis*, UNHCR (2020). UN Human Rights Committee Decision on Climate Change Is a Wake-Up Call, According to UNHCR, https://www.unhcr.org/news/briefing/2020/1/5e2ab8ae4/un-human-rights-committee-decision-climate-change-wake-up-call-according.html. Accessed 10 February 2020; Sinclair-Blakemore (2020). Teitiota v New Zealand: A Step Forward in the Protection of Climate Refugees Under International Human Rights Law?, *Oxford Human Rights Hub Blog*. https://ohrh.law.ox.ac.uk/teitiota-v-new-zealand-a-step-forward-in-the-protection-of-climate-refugees-under-international-human-rights-law/. Accessed 13 March 2020; Amnesty International (2020). UN Landmark Case for People Displaced by Climate Change. https://www.amnesty.org/en/latest/news/2020/01/un-landmark-case-for-people-displaced-by-climate-change/. Accessed 16 February 2020.

[52] See for instance *Teitiota v The Chief Executive of the Ministry of Business, Innovation and Employment*, decision of the High Court of New Zealand no. [2013] NZHC 3125 of 26 November 2013, para. 63.

rise was not adequately supported by convincing evidence on the particular risks he and his family were individually facing.[53] Furthermore, the Committee found that New Zealand's authorities had sufficiently considered his individual and particular situation.[54] As a result, the Committee "consider[ed] that [...] the information made available to it does not demonstrate that the conduct of [New Zealand's] judicial proceedings in the case was clearly arbitrary or amounted to a manifest error or denial of justice, or that the courts otherwise violated their obligation of independence and impartiality".[55]

In a nutshell, Mr. Teitiota was not able to produce convincing evidence of an *imminent risk* to his life, a threshold that is to be met to configure a violation of Article 6 of the ICCPR. In the Committee's view, the fact that the direst effects of sea-level rise will occur in ten/fifteen years, gives to the government of Kiribati a sufficient timespan to take the measures that are necessary to avoid its most harmful consequences to its citizens, Mr. Teitiota included.[56]

Notwithstanding this conclusion, the Committee has also made a clear and important leap forward in view of the possible recognition of international protection rights for individuals who suffer from adverse and extreme climatic conditions in their origin countries. It stressed that without a robust national and international effort, the effects of climate change in these countries may expose individuals to violations of their rights under Articles 6 and 7 of ICCPR, thereby activating *non refoulement* obligations: a state should hence not repatriate individuals where the effects of climate change expose them to life-threatening risks or to risks of undergoing inhuman or degrading treatment. Even though the particular condition of Mr. Teitiota and his family did not configure such a risk at present, the Committee pointed out that state parties are requested to consider the impact on the enjoyment of the right to life of the effects of climate change, be they of gradual or sudden-onset nature, since both

[53] For a dissenting opinion on the specific issue of water availability in Tarawa, please refer to the Individual opinion of Committee member Vasilka Sancin. UN Human Rights Committee (HRC) (2020). *Ioane Teitiota v. New Zealand* (cit.), Annex 1.

[54] UN Human Rights Committee (HRC) (2020). *Ioane Teitiota v. New Zealand* (cit.), para. 9.6–10.

[55] Ibid., para. 9.12.

[56] Ibid.

can push individuals to cross borders, abandon their countries, and seek refuge abroad.[57]

In a very interesting way then, the Committee has opened up to the possibility for forced climate migrants of obtaining protection in a third country, clarifying that an individual cannot be repatriated if climate change has led to uninhabitable life conditions in the country of origin. With particular reference to the situation of a country like Kiribati, it furthermore underlined that the possibility that an entire country is submerged represents an extreme risk that may render life conditions incompatible with the right to life even before the complete "sinking" of the country. In its whole reasoning, the Committee took account of General Comment no. 31 on the nature of the obligations imposed on state parties to the ICCPR and of the obligations that states bear in guaranteeing the right to life, specifying that states must not extradite (*refouler*) from their territories an individual that is under threat of suffering an irreparable harm in his or her country of origin.[58] In consideration of General Comment no. 36, it then also noted that environmental degradation and climate change represent one of the most serious threats to present and future generations in this regard.[59] Thus, even though Mr. Teitiota's appeal was rejected, it comes as a decisive clue in clarifying how *non refoulement* obligations should apply in cases of international migrations forced by climate change.

Right to Health, Access to Livelihoods, and Climate Change

Intimately connected to the right to life is of course the right to health, whose respect must be granted pursuant to Article 12 of the Covenant on Economic, Social and Cultural Rights (ICESCR). States are required to undertake all actions aimed at ensuring that their citizens live in conditions that guarantee health levels, both from a mental and physical point of view, that are as high as possible. Global warming and its effects can have serious negative implications on this right. In fact, increasing

[57] Ibid., para. 9.11.

[58] Ibid., para. 9.3.

[59] Ibid., para. 9.4. Refer in this regard to UN Human Rights Committee (HRC) (2020). General comment no. 36: Article 6 (Right to Life) (op. cit.), para. 62.

global temperatures lead to a higher possibility of the spreading of tropical diseases, above all malaria, in traditionally temperate climate zones. Furthermore, the melting of permafrost can release bacteria and diseases that have literally frozen since from time immemorial, and whose effects on currently living human populations are to be completely discovered.[60] Alongside this, the increase and intensity of extreme weather events can clearly exacerbate situations that have a negative impact on human health: just consider for instance all physical injuries that can occur in the event of rapid-onset disasters and the possibility of spreading diseases like cholera in periods subsequent to disasters such as floods or cyclones.

The right to health obviously implies the need to have free access to basic means of sustenance, such as food, and water. It also equally implies the necessary existence of environmental and socio-economic conditions capable of making a specific region suitable to healthy and acceptable living conditions.[61] The importance of the access to livelihoods in relation to *non refoulement* is also witnessed by the ECHR.[62] When called to decide on asylum seeker cases, it assesses the claimant's ability to meet his or her most basic needs, such as food, hygiene, and shelter, his/her vulnerability to ill-treatment, and the prospect of his/her situation improving within a reasonable time frame.[63]

The right to food and the right to water are thus clearly relevant and complementary to the right to health.[64] Pursuant to the ICESCR, the right to food is an integral part of the right to an adequate standard

[60] See Goudarzi (2016). As Earth Warms, the Diseases That May Lie Within Permafrost Become a Bigger Worry *Scientific American*. https://www.scientificamerican.com/article/as-earth-warms-the-diseases-that-may-lie-within-permafrost-become-a-bigger-worry/. Accessed 12 May 2019.

[61] In this sense, see UN Committee on Economic, Social and Cultural Rights (CESCR) (2000), CESCR General Comment No. 14: The Right to the Highest Attainable Standard of Health (Art. 12). https://www.refworld.org/docid/4538838d0.html. Accessed 14 June 2019.

[62] Please refer for a detailed analysis to Chapter 11 and, more specifically, European Court of Human Rights, *MSS v. Belgium & Greece*, Application no. 30696/09, 21 January 2011, para. 252–254. See also European Court of Human Rights, *Sufi & Elmi v. United Kingdom*, Applications no. 8319/07 and 11449/07, 28 June 2011, para. 282.

[63] Poon (2018) (op. cit.). pp. 162–163.

[64] For more information, see Franca (2016). Climate Change and Interdependent Human Rights to Food, Water, and Health: The Contest Between Harmony and Invention. In Quirico & Boumghar (Eds.), *Climate Change and Human Rights: An International and Comparative Law Perspective*. New York: Routledge, p. 89 ff.

of living, as governed by Article 11. This right is also expressed—and despite the long-standing difficulties in seeing it implemented and respected throughout the world[65]—in the universal right of freedom from hunger. Despite the more or less praiseworthy efforts of the international community, which set in 2015, with the adoption of the 2030 Sustainable Development Goals,[66] the goal of halving the number of people suffering from hunger worldwide, there is no doubt that climate change may worsen hunger-related concerns in many parts of the globe, with particularly significant impacts in arid areas subject to desertification in sub-Saharan Africa. In this sense, the words pronounced in 2008 by then UN Special Rapporteur on the right to food, Jean Ziegler, are clear: "half of the world's hungry people depend on their survival on lands which are inherently poor and which may be becoming less fertile and less productive as a result of the impact of repeated droughts, climate change and unsustainable land use".[67] Global warming also has decisive effects on the distribution of water resources and, consequently, on the possibility to access water. On the one hand, the increase in floods and torrential rains can cause an "excess" in some regions, while on the other, desertification and increasingly powerful drought, together with the decrease in the volume of glaciers, produce exactly the opposite effect.[68]

Although ICESCR's Article 11 does not make direct reference to the existence of an individual right to water, it is at any rate obvious that the exercise of the rights to life, health, and an adequate standard of living crucially depend on access to water. It is not surprising, then, that the UN Committee on Economic, Social and Cultural Rights (CESCR) and the UN General Assembly have expressed themselves clearly in this

[65] Sobering to recall that approximately more than 800 million worldwide people suffer from malnutrition.

[66] UN General Assembly Resolution (2015). Transforming Our World: The 2030 Agenda for Sustainable Development (A/RES/70/1).

[67] UN Human Rights Council (2008). Report of the Special Rapporteur on the Right to Food, Jean Ziegler. Para. 5. https://www.refworld.org/docid/47c3dbe82.html. Accessed 15 January 2020.

[68] See UN-Water, Climate Change adaptation Is Mainly About Water. https://sustainabledevelopment.un.org/content/documents/UNWclimatechange_EN.pdf. See also UN-Water (2017). Climate Change Adaptation. The Pivotal Role of Water. UN-Water (2019). Policy Brief CLIMATE CHANGE AND WATER. Final Draft. https://www.unwater.org/publications/un-water-policy-brief-on-climate-change-and-water/. Accessed 20 March 2020.

regard, specifying how the right to water consists in the possibility of accessing sufficient, healthy, and affordable water resources.[69] Obviously, given the complexity of climate change, the variation in the distribution of water resources will increasingly influence other factors, such as access to food, soil degradation, population pressure, and the ability of affected communities to adapt current changes, namely their degree of resilience. It therefore becomes appropriate to look at the interplay between climate change and human rights with a broad and holistic approach, and not to focus on one single fundamental right per time.

Indeed, climate change is and will increasingly alter several ecological balances on which human societies have based on their development—on more or less stable bases–for about 10,000 years, or since the beginning of the Holocene. The alteration of these balances will depend on the future magnitude of climate change itself and therefore largely on the capacity and willingness of humankind to face these challenges. In the most vulnerable areas and in the poorest countries of the world, where it is already difficult to access food and water and, consequently, and where the rights to life and health are already jeopardized, climate and ecological conditions will inevitably exacerbate fragile situations, thus contributing to an increase in migration flows worldwide.

It is unquestionable that global warming has an impact on the exercise of fundamental rights,[70] but it remains to be seen to what extent it does and above all how the international community will act to face its most negative effects both today, and more so, tomorrow.

[69] Cf. UN Committee on Economic, Social and Cultural Rights (CESCR) (2003). General Comment No. 15: The Right to Water (Arts. 11 and 12 of the Covenant). https://www.refworld.org/docid/4538838d11.html. Accessed 13 September 2019. Refer also to UN General Assembly Resolution (2010). The Human Right to Water and Sanitation, UN Doc. A/RES/64/292.

[70] See in this sense Knox (2015) (op. cit.). See also Decision 1/CP.16, The Cancun Agreements: Outcome of the Work of the Ad Hoc Working Group on Long-term Cooperative Action under the Convention, UNFCCC COP 16.

CHAPTER 9

Why States Should Protect Forced Climate Migrants

We have so far outlined and analyzed all the building blocks constituting the rationale for the international community's obligation to provide dedicated protection mechanisms for the category of forced climate migrants. Let us briefly review these.

The definitional analysis presented in Chapter 5 concluded that a forced climate migrant is a person compelled to leave his or her country because this has become or is about to become uninhabitable due to serious environmental degradations caused by climate change. By linking this definition to what we discussed in the last chapter on fundamental human rights, we can consider a country uninhabitable when it is unable to cope with climate change-driven environmental changes that seriously challenge the possibility on the whole national territory to access basic livelihoods such as food and water, an eventuality that negatively affects health standards and even the life of the individuals involved.

Understood as such, forced climate migrants are human beings subject to deprivations of universal fundamental rights, which in turn force them to leave their country. Since these deprivations are ultimately caused by climate change, a phenomenon for which it is possible to identify a human origin and agency, there should be an international duty to predispose protection mechanisms for them. If a person is not able to rely anymore on his or her country because of global warming, namely

G. Sciaccaluga, *International Law and the Protection of "Climate Refugees"*,
https://doi.org/10.1007/978-3-030-52402-9_9

because of something created and exacerbated collectively by the international community, shouldn't the same community act to protect this person? If international law protects political refugees because they cannot rely on their own country—that is directly or indirectly persecuting them—shouldn't international law protect even more so forced climate migrants when they cannot live in their own countries any longer because of a phenomenon for which the entire international community is responsible? With a political refugee, the community must intervene to protect that person because he or she is persecuted by his/her own state, thus not being able to avail him/herself of the protection of that state. We can rely on the same reasoning to argue that forced climate migrants should be protected by international law too. The international community should envisage protection systems for this new category of forced migrants when "individuals are no longer able to avail themselves of the protection of their national state. In the case of climate displaced persons, the country of origin is unable to protect their citizens either because it is overwhelmed with severe weather events, lacks resources or because the state itself is disappearing".[1]

When we consider a forced climate migrant, the impossibility to exercise basic human rights in the country of origin (whose effects in terms of suffering are comparable to persecution) is determined by the collectivity of states. Shouldn't then their community step into protect that person because she has become unable to avail herself of the protection of her state?

Still, at the United Nations level, in the words of the Human Rights Council's and the OHCHR's resolutions, as well as under the UNFCCC for example, there is a clear intent to avoid explicit references to "human rights violations" deriving from states' failures to fulfill their obligations to mitigate and adapt to climate change. These discussion *fora* prefer to talk about "negative effects and implications" on the exercise of fundamental rights, with the specific intent to relieve states of direct responsibilities.[2] However, in view of the ever-growing body of univocal scientific consensus, as well as the gradual worsening of climate change with its

[1] Atapattu (2018). A New Category of Refugees? "Climate Refugees" and Gaping Hole in International Law. In Behrman & Kent (Eds.), *"Climate Refugees" Beyond the Legal Impasse?* New York: Routledge, pp. 41 and 48.

[2] See in this sense Francioni & Quirico (2016). Untying the Gordian Knot. In Quirico & Boumghar (Eds.), *Climate Change and Human Rights*. New York: Routledge, p. 151.

ensuing disastrous environmental damages, this approach will become more and more indefensible over time. Under this perspective, climate change (but in reality, more upstream, those climate-altering activities allowed or promoted by states) can already be considered today—and will increasingly be so tomorrow—as the cause of potential infringements of human rights that are universal and mandatory under international law.

Following this line of argument, the last question to be discussed in-depth concerns the obligations that states have in protecting individual fundamental rights. We already saw which human rights can be negatively affected by climate change. Here, the focus moves to the consequent obligations that weigh internationally on states, with a particular regard to the actions that cause and exacerbate climate change and, thus, to their greenhouse gas emissions. Generally, in relation to fundamental human rights, a state has the obligation of respect and protection,[3] bearing consequently both negative and positive obligations on different levels. The obligation to respect reflects, above all, the prohibition to interfere in the exercise by an individual of his/her right.[4] As a proof of its centrality, the obligation

They moreover argue that "effectively grappling with climate change requires a radical re-conceptualization of human rights and environmental damage. Such re-conceptualization is necessary to the extent that climate change is a global problem for humanity as a whole". See also Stoutenberg (2013). When Do States Disappear?: Thresholds of Effective Statehood and the Continued Recognition of "Deterritorialized" Island States. In Gerrard & Wannier (Eds.), *Threatened Island Nations: Legal Implications of Rising Seas and a Changing Climate*. New York: Cambridge University Press, pp. 75–76: "[T]here is a difference between saying that a right has been affected (i.e., that its exercise has been factually impaired) and that it has been violated through a breach attributable to another State".

[3] The UN Committee on Economic, Social and Cultural Rights has repeatedly held this point, starting from 1999. See UN Committee on Economic, Social and Cultural Rights (CESCR) (1999). General Comment No. 12: The Right to Adequate Food (Art. 11 of the Covenant). https://www.refworld.org/docid/4538838c11.html. Accessed 30 September 2019.

[4] See on similar terms: Skogly (2006). *Beyond National Borders: States Human Rights Obligations in Development Cooperation*. Antwerpen-Oxford: Intersentia, p. 68. See also the Ogoniland case (cit.) para. 45: "At a primary level, the obligation to respect entails that the State should refrain from interfering in the enjoyment of all fundamental rights; it should respect right-holders, their freedoms, autonomy, resources, and liberty of their action".

to protect is for instance defined as "*a fundamental cross-cutting obligation in a general provision covering all Convention rights*"[5] within the American Convention on Human Rights. In the context of the ICCPR, it is articulated in Article 2.1, according to which: "*each of the states parties to the present Covenant undertakes to respect and guarantee to all individuals who are in its territory and are under its jurisdiction the rights recognized in the present Covenant […]*". The obligation to respect also finds its formulation in the ICESCR.[6]

Normally, the obligation to respect takes the form of a negative obligation, under which a state is limited in the exercise of its sovereign power when this is likely to threaten the exercise of fundamental rights of individuals. It imposes both an obligation of result and conduct, by virtue of which states must abstain from directly committing a violation of the right involved and impede infringements of the same by third parties. With regard to climate change, it may be possible to invoke an obligation of noninterference in the exercise of human rights under which a state should not worsen or accelerate global warming, so as not to undermine the exercise of the fundamental rights of its citizens[7] as well as of other states. In other words, "given that climate change is generally accepted or accelerated by authorizing [GHG] emissions, this obligation would result in such actions as unsuccessful as they result in harmful impact to the realization of rights".[8] Such an argument can be made even more clearly, when it comes to the obligations toward the populations of insular micro-states at risk of submersion, given their particular condition and

[5] American Convention on Human Rights, adopted 22 November 1969, entered into force 18 July 1978, Article 1.

[6] More precisely, in relation to the obligation of non-discrimination. Cfr. UN Committee on Economic, Social and Cultural Rights (CESCR) (1990). General Comment No. 3: The Nature of States Parties' Obligations (Art. 2, Para. 1, of the Covenant), p. 86. https://www.refworld.org/docid/4538838e10.html. Accessed 28 March 2019.

[7] See in this sense The Hague District Court (2015) in occasion of the *Urgenda case* (Rechtbank den Haag, 24 June 2015, C/09/456689 / HA ZA 13-1396). https://uitspraken.rechtspraak.nl/inziendocument?id=ECLI:NL:RBDHA:2015:7196. Accessed 12 May 2019. See in this sense also: McInerney-Lankford (2013). Human Rights and Climate Change. In Gerrard & Wannier (Eds.), *Threatened Island Nations: Legal Implications of Rising Seas and a Changing Climate*. New York: Cambridge University Press, p 236: "Positive obligations require that States act to protect their citizens from the impacts of climate change, regardless of whether the harms of climate change result from their acts or omissions and regardless of whether the State is itself to blame for climate change".

[8] McInerney-Lankford (2013) (op. cit.), p. 212.

vulnerability. Poon argues in this regard that the inhabitants of sinking micro-states should not be repatriated when they leave their state in case their life or freedom there is seriously threatened.[9]

Complementary to the obligation to respect, the obligation to protect requires states to regulate the behavior of third parties, thus mainly of private subjects, in order to protect the human rights of those who are under their jurisdictions. This obligation usually takes the form of a duty to establish, through acts of law, regulation, prevention, investigation, sanction, and compensation measures.[10] The concept of due diligence, which we already discussed, becomes relevant once again. When a state fails to regulate the activity of the private actors operating on its territory, leaving them free to interfere with the exercise of human rights of individuals, it is disrespecting its obligation to protect. In the context of climate change, it can be assumed that states should take the measures that are necessary to protect people under their jurisdiction from the negative effects of global warming that can create environmental conditions in which the effective exercise of certain fundamental rights (life, health, access to means of support) cannot be granted. In applying due diligence, one could thus consider that the obligation to protect also involves the need to regulate and limit the conduct of private actors that cause or exacerbate climate change. In practical terms, this obligation should consist in establishing a legal framework to regulate and reduce national greenhouse gas emissions and to implement policies and plans to adapt to the negative effects of global warming.

In all this, a further important question to deal with touches the extraterritorial scope of these non-interference obligations in the exercise of human rights. Under an extraterritorial human rights violation perspective, the responsible states would essentially be those that, once aware of the effect of their greenhouse gas emissions, have not undertaken serious and proportionately adequate policies to cut their national emissions even though under an obligation to do so imposed by general

[9] Poon (2018). Drawing upon International Refugee Law. The Precautionary Approach to Protecting Climate Change-Displaced Persons. In Behrman & Kent (Eds.), *"Climate Refugees" Beyond the Legal Impasse?* New York: Routledge, p. 163.

[10] In this regard, please refer again to the *Ogoni case* (cit.), para. 46: "At a secondary level, the State is obliged to protect right-holders against other subjects by legislation and provision of effective remedies. This obligation requires the State to take measures to protect beneficiaries of the protected rights against political, economic and social interferences".

and/or particular international law.[11] The most important among these actors are, as broadly known and repeated in every interested field, industrialized and developed states and, today, certainly also a country like China for example. These are countries where the negative effects of climate change are surely felt, but, interestingly, where national resilience capabilities allow to contain their adverse effects on the exercise of the fundamental rights of their citizens within limits of tolerability.[12] Rather, the actions (or inactions) of these countries in terms of greenhouse gas emissions have a more decisive negative impact on individuals and populations of other states, that are geographically more vulnerable to and historically less responsible for the greenhouse effect.

It is hence interesting to look at the obligations to which exceedingly greenhouse gas emitting states are or should be subject to also extraterritorially.[13] Our argument here is not to maintain that large polluting countries, being the most responsible for anthropogenic climate change, are solely responsible for climate-related human rights infringements and, in turn, for the protection of forced climate migrants, but rather to understand how extraterritorial human rights obligations can and should affect climate-altering state conducts. In this regard, it is commonly assumed that a state should not affect the exercise of human rights outside its own territory. This obligation also applies to the prohibition—that is relevant for our purposes, especially with regard to insular micro-states—of

[11] In this sense and in application of the principle of due diligence appear relevant the recent studies according to which some major fossil fuel companies have deliberately hidden the contribution of their activities to the phenomenon of climate change. See Supran & Oreskes (2017). Assessing ExxonMobil's Climate Change Communications (1977–2014). *Environmental Research Letters*, 12. https://iopscience.iop.org/article/10.1088/1748-9326/aa815f/pdf. Accessed 13 March 2019.

[12] The *Urgenda case* in the Netherlands shows that the citizens of these countries also have started to claim for the respect of their human rights against internal economic activities that contribute to climate change (for more information, see *sub* note 279).

[13] See Meron (1995). Extraterritoriality of Human Rights Treaties. *American Journal of International Law*, 89, pp. 78 ff. Skogly (2006). *Beyond National Borders: States Human Rights Obligations in Development Cooperation*. Antwerpen-Oxford: Intersentia. Craven (2007). The Violence of Dispossession: Extraterritoriality and Economic, Social and Cultural Rights. In Baderin & McCorquodale (Eds.), *Economic Social and Cultural Rights in Action*. Oxford: Oxford University Press, p. 71, where the author claims, with regard to the ICESCR that "while a State's primary responsibility is to the local populace; each State is also required to ensure that it does not undermine the enjoyment of rights of those in foreign territory".

undermining the ability of a third state to fulfill its obligation to respect and protect human rights in its territory. In support of this argument, it is also possible to invoke an extensive interpretation of the *sic utere tuo* principle, as well as what established by the Stockholm and Rio Declarations on the environment. Particularly relevant is Article 21 of the 1972 Stockholm Declaration, according to which "*states have, in accordance with the UN Charter and the principles of international law, the right to exploit their own resources according to their environmental policies, and the responsibility to ensure that their activities are within the limits of national jurisdiction*". More important however is the endorsement of the International Court of Justice[14] and of appointed human rights bodies[15] to the existence of extraterritorial obligations of states, especially in cases where an adequate control over their activities has not been exercised. Furthermore, with particular reference to the right to food, the UN Committee on Economic, Social and Cultural Rights noted that "[t]he extraterritorial obligation to respect the right to food requires states to 'do no harm' [...]. It is necessary to ensure that their policies and measures do not lead to violations of the right to food for people living in other countries".[16] The *sic utere tuo* principle therefore, even though not directly applying to the relationship between a state conduct and human rights infringements, is here used to underline the responsibility of states in the creation of planetary, climate change-induced environmental degradation. Clearly, the principle's applicability and effectivity are normally limited to easily

[14] International Court of Justice, *Armed Activities on the Territory of the Congo (Democratic Republic of the Congo v. Uganda)*, I.C.J. Reports 2005, p. 168, para. 220; *Legal Consequences of the Construction of a Wall in the Occupied Palestinian Territory*, Advisory Opinion, I.C.J. Reports 2004, p. 136, para. 111.

[15] See UN Human Rights Committee (HRC), *Sergio Ruben Lopez Burgos v. Uruguay*, Communication No. R.12/15, para. 176 (UN Doc. a/36/40, Supp. No. 40, 1981). See also Inter-American Court of Human Rights, *Coard* et al. *v. United States*, case 19.951, Report No. 109/99, 1999. See also what maintained by the UN Human Rights Committee in particular with reference to Article 2 of ICCPR: "a State party must respect and ensure the rights laid down in the Covenant to anyone within the power or effective control of that State party, even if not situated within the territory of the State party", in UN Human Rights Committee (HRC) (2004). General Comment No. 31 [80], The Nature of the General Legal Obligation Imposed on States Parties to the Covenant, para. 10. https://www.refworld.org/docid/478b26ae2.html. Accessed 20 March 2020.

[16] UN Committee on Economic, Social and Cultural Rights (CESCR) (1999). General Comment No. 12: The Right to Adequate Food (Art. 11 of the Covenant), para. 19. https://www.refworld.org/docid/4538838c11.html. Accessed 23 September 2019.

identifiable transboundary pollution cases, but it is at the same time a useful and thought-provoking legal tool thanks to which it is possible to link climate-altering state conducts to human rights infringements. Leveraging on the *sic utere tuo* principle and extraterritorial environmental degradation caused by climate change makes hence a further point as to why states should not permit excessively climate-altering policies when these have the potential of impeding the enjoyment of fundamental rights of individuals, both within and beyond their domestic borders.

In this line of reasoning, it is quite clear then that human rights obligations can be invoked to impose onto states certain conducts regarding their climate policies, particularly with a view to protecting the rights of those individuals who are under their jurisdiction. And this especially if the argument of the case law concerning the *Urgenda* case—where the Dutch government was condemned for not having implemented greenhouse gas emission reduction measures deemed sufficient to fulfill its obligations to protect the rights of its citizens[17]—found over the years its application and diffusion in more than one legal order, as it looks today quite probable.[18] On the other hand, the extraterritorial reach of human

[17] Please refer in this regard to Sciaccaluga (2015). La Corte Distrettuale dell'Aja e la lotta ai cambiamenti climatici: valori e limiti della politica climatica dell'Unione europea. *Eurojus.it Rivista*. http://rivista.eurojus.it/la-corte-distrettuale-dellaja-e-la-lotta-ai-cambiamenti-climatici-valori-e-limiti-della-politica-climatica-dellunione-europea/. Accessed 13 April 2019.

[18] In various legal systems, individuals have started to appeal against their national governments with the intention of seeing them condemned by the judicial system for the inadequacy of their climate policies. At the same time, some public entities have started to appeal against major fossil fuel companies, allegedly accused of having undermined and hidden their role in contributing to global warming. See for instance, Farand (2017, April 1). Nine Year Old Girl Files Lawsuit Against Indian Government over Failure to Take Ambitious Climate Action. *The Independent*. https://www.independent.co.uk/environment/nine-ridhima-pandey-court-case-indian-government-climate-change-uttarakhand-a7661971.html. Accessed 23 January 2020. Wang (2017, October 5). Philippines Climate Case Could Find Fossil Fuel Companies Violate Human Rights. *Climate Liability News*. https://www.climateliabilitynews.org/2017/10/05/philippines-climate-change-human-rights/. Accessed 13 May 2019. DiChristopher (2018, July 19). Judge Throws Out New York City's Climate Change Lawsuit Against Five Major Oil Companies. *CNBC*. https://www.cnbc.com/2018/07/19/judge-tosses-nycs-climate-change-lawsuit-against-5-big-oil-companies.html. Accessed 20 July 2019. Kaminski (2019, December 9). Carbon Majors Can Be Held Liable for Human Rights Violations, Philippines Commission Rules. *Climate Liability News*. https://www.climateliabilitynews.org/2019/12/09/philippines-human-rights-climate-change-2/. Accessed 20 March 2020.

rights obligations, while endowed with its own specific *raison d'être* and international consideration, is not very incisive on a practical level due to the ineffective enforceability of human rights law. This is primarily because human rights treaties tend to impose, by their very nature, vertical obligations that link governments to their citizens, with jurisdictional protection mechanisms usually limited to the relations between these subjects.[19] On the contrary, climate change requires a different approach, since it inherently also opens up to relations between a state and individuals that are outside its territory and jurisdiction. Consequently, despite the reasoning carried out so far on the extraterritorial scope of state obligations in the matter of human rights, their application and effectiveness remains limited and disputed today.[20]

Nonetheless, it is reasonable to claim that such obligations could (and should) have both a direct and indirect impact on the implementation of the policies aimed at combating and managing climate change. More precisely, they can work as a benchmark for the content and objectives of these policies, as well as—more generally—with a view to environmental protection.[21] If this is the case, we can argue that the international human rights regime imposes as a minimum a do no harm obligation under which state actions and omissions with respect to the problem of global warming must at least not decisively contribute to the impossibility to enjoy of basic human rights nationally and, to a lesser extent, abroad.[22] Again, the very particular rights and interests, both on a collective and individual level, of insular micro-states whose "deterritorialization" due to climate change seems difficult to avoid, come at the forefront. Of course, technical and scientific obstacles emerge: when and beyond which threshold is a state conduct to be considered irrespective of human rights domestically and—decidedly more complex—also extraterritorially? If we

[19] Henkin (1990). *The Age of Rights*. New York: Columbia University Press (cit. in McInerney-Lankford 2013) (op. cit.), p. 220: "A variety of reasons may be put forward [to explain the limitation of human rights extraterritorial obligations]: 'because a State may not ordinarily exercise authority in the territory of another State, there is no obligation upon it to act to ensure respect for human rights there, whether by the government of that State or by private persons'".

[20] Ibid. (also for further reference), p. 231.

[21] Cf. UN Human Rights Council (2011). Resolution 16/11, Human Rights and the Environment (UN Doc. A/HCR/RES/16/11), p. 2.

[22] McInerney-Lankford (2013) (op. cit.), pp. 235 and 236.

follow the *Urgenda* case reasoning, the determination of this threshold ultimately depends on the national judiciary system that, in application of both national and international law and in the light of available scientific knowledge on climate change, is in the position of evaluating the amount of national greenhouse gas emissions in terms of their impact on human rights, especially at the domestic level.

The precautionary principle can become relevant again here, also with a view to protecting forced climate migrants. As seen in Chapter 6, this principle applies in environmental protection matters, requesting states to act and to protect the environment even before a causal link between a certain activity and an occurring environmental degradation has been established by absolute clear scientific evidence. If we merge the rationale of the precautionary principle with climate change, human rights, and *non-refoulement* considerations, it becomes possible to argue that states should undertake precautionary measures to prevent global warming, bearing also a corresponding duty to assist those who flee from its effects.[23]

Anthropogenic climate change is a cause that can "create" an environmental condition in which certain mandatory and inalienable fundamental rights cannot be exercised in some countries, compelling their inhabitants to emigrate abroad in search for assistance. If science shows that climate change is also determined by human activities for which states can be directly or indirectly accountable under international law, and if "this" climate change creates or exacerbates environmental situations which hamper universal and non-derogable fundamental rights, infringements in presence of which a new legal category of individuals, i.e., "forced climate migrants", are forced to leave their country of origin, then the international community has the obligation to protect this category through mechanisms, rules, and instruments of international protection.

When considering climate change-forced migration, we saw that a hypothetical responsibility could be invoked where states and the whole international community keep allowing or carrying out economic activities that decisively affect global warming, which determines, in some areas

[23] Poon (2018) (op. cit.), p. 159. Cf. also UN Human Rights Committee (HRC) (2020). *Ioane Teitiota v. New Zealand* (cit.), para. 9.11, when it states that "given the risk of an entire country becoming submerged under water is such an extreme risk, the conditions of life in such a country may become incompatible with the right to life with dignity *before* the risk is realized" (emphasis added).

or even in entire states, environmental situations that make it is impossible to enjoy fundamental human rights. Thus, when individuals living there are denied the possibility of staying in their own state, a right to assistance and protection abroad is to be recognized. This is because, in essence, a forced climate migrant can no longer—mainly for responsibilities not attributable to his/her own state of origin, but rather to all states together, although obviously with differentiated responsibilities—enjoy the relationship of trust, collaboration, and assistance he or she should normally enjoy with the country of origin. International law grants traditional political refugees international protection when they can no longer exercise their rights in the political and territorial space of the origin state. In the hypothetical "forced climate migrants' law", an equal or at least similar level of protection should be granted on the very same basis: hence, when these people are denied the possibility of exercising fundamental rights in their territorial and political space, despite the different causes behind this impossibility: political persecution *v.* climate change-driven environmental degradation.

Conclusion

If science has some degree of validity, the causal relationship between human activity and global warming should be taken for granted. As seen throughout this book, this certainty is clear—at least in theory—also under international law, where the UNFCCC and its 197 parties agree on the human origins of climate change. This even more so in light of the 2015 Paris Agreement, through which the vast majority of states and the entire international community decided to undertake climate obligations, recognizing the role that certain economic activities have on global climate patterns.

If so, it is becoming increasingly difficult to argue that all states, even though bearing differentiated responsibilities, are not internationally responsible when they fail to undertake adequate greenhouse gas emission reduction/containment measures. Obviously, these must be proportional to the national interests and capabilities of every single state, but equally important is the need to reach the common goal set out in Paris of safeguarding the stability and tolerability of our shared climate system. We saw in fact the reasons why it is possible to conclude that this objective imposes *erga omnes* obligations, given the nature, importance, and transversality of the "good" they intend to protect: the climate balance in which the international community has thrived and developed for centuries. We also saw that climate-altering state actions can jeopardize the exercise of certain mandatory and inalienable fundamental rights. The extraterritorial scope of international human rights law is, above all from an enforceability point of view, highly deficient. Nevertheless, we can argue that those who are most vulnerable to the negative effects of global warming, being obliged to leave their countries, should be protected through instruments of international law in the name of the protection of inalienable and non-derogable fundamental rights. The (albeit still timid) attempts by the international climate change regime aimed at including climate-induced migration and displacement into the list of issues that have to be regulated can be considered as the embryo of a future regulatory framework which aims at protecting forced climate migrants. Intergovernmental efforts such as the Nansen Initiative and the Platform on Disaster Displacement go in this direction as well, alongside the soft law development witnessed by the adoption of the 2018 Global Compacts on Migration and Refugees.

Coming to a conclusion, when an individual, unable to exercise fundamental rights, is forced to leave his or her national territory because this has become (or is about to become) uninhabitable due to global warming, a right to international protection abroad must be recognized to this individual given the anthropogenic origin of climate change and the mandatory scope of the fundamental rights put at risk. In the domain of refugee law, international protection is guaranteed when someone is barred from exercising his fundamental rights at home because his/her own government (in)directly persecutes him or her. When we consider forced climate migrants, there are very similar needs and underlying characteristics.[24] The fact that the subject primarily responsible for the impossibility of exercising one's rights at home here is not the country of origin, but, rather, the international community as a whole, does not represent a sufficient factor for denying a protection regime to forced climate migrants. In fact, the opposite is to be argued. If the international community and its laws protect a person where she is not sufficiently

[24] In this sense, please refer to the theses supported, *ex multis*, by McAdam, according to which it would be necessary today to set up protection mechanisms for forced migrants according to their needs and no longer on the basis of the reasons underlying their difficult situation, persecution, or danger. See McAdam (2012). *Climate Change, Forced Migration, and International Law*. Oxford: Oxford University Press, Chapter 9.

protected by her country of origin, then a similar protection should also be guaranteed when individuals are not sufficiently protected by their country of origin due to the conduct of that same and whole international community.

In a nutshell, we are today witnessing the progressive emergence of a duty of international protection for forced climate migrants. In the past, the eighteenth century brought to the consolidation of the international moral principle proclaiming the existence of a state duty to grant protection to political refugees.[25] A similar moral principle imposing a duty to protect forced climate migrants, identified *only* as those who are forced to leave their country of origin due to serious environmental degradation caused by climate change, could—and maybe also should—similarly find room in the doctrine and practice of the twenty-first century.

[25] Cf. Lenzerini (2009). *Asilo e diritti umani- Evoluzione del diritto d'asilo nel diritto internazionale*. Milan: Giuffré, p. 31.

PART III

The Embryos of Forced Climate Migrants' Law

Introduction

Over, the years, in response to the challenges posed by climate-related migration, many initiatives and programs have been proposed, both from the academic and institutional worlds. Among these, we may recall for instance the Nansen Initiative and its Protection Agenda, the Peninsula Principles on Climate Displacement, or the Draft Convention on the International Status of Environmentally Displaced Persons.

Once the inapplicability of current legal frameworks to the category of "climate refugees" was acknowledged, international law experts started studying the possibility to create *ad hoc* international law instruments and/or to extend the scope of application of potentially appropriate legal systems, such as the 1951 Refugee Convention or complementary protection frameworks for example. Synthetically, whereas maximalist scholars have mainly stressed the need to create new, dedicated hard law instruments, minimalists have criticized the political (un)feasibility of this option. In their opinion, the scientific uncertainties and the political sensitivity that surround the phenomenon make it improbable that states would be prone to successfully gather with the objective of creating/extending international protection rules so as to cover those who are displaced by climate change.

This author agrees with this position: the chances for the international community to create incisive *ad hoc* rules or instruments appear far too remote, at least in the short/medium run. The gradual development of soft law instruments and the evolution of complementary protection systems may on the contrary represent workable alternatives to provide forced climate migrants with international protection. In the previous chapters, we saw the reasons why international law should equip itself with protection mechanisms for this category. In the name of the protection of non-derogable human rights and in respect of international climate change law, the international community should put in place measures to protect those who are effectively forced to leave the entirety of their national territory due to environmental degradations caused by climate change. If such a duty is on the verge of being established, we still need to see how (and under which frameworks) it may effectively concretize in the future. We have therefore to understand where there is room for states to address climate-induced migrations and to predispose protection mechanisms for forced climate migrants.

In this sense, in the next two chapters we are going to look at those frameworks within which a future "forced climate migrants law" is been progressively built. In fact, its embryos are already there: the progress made under the UNFCCC in the last years and the adoption of the Refugee and Migration Compacts in 2018 go in this direction and witness the international's community gradual development toward the establishment of a multifaceted system dedicated to the management of the "climate refugees" phenomenon. Chapter 10 is dedicated to the analysis of the developments recently occurred within the UNFCCC in the field of climate migration and with the adoption of the Global Compacts on Migration and Refugees, that have for the first time explicitly and thoroughly recognized the role that environmental degradation and global warming have in triggering cross-border migration. If the UNFCCC and the two Global Compacts represent therefore the frameworks under which it is possible to develop protection guidelines and regulations for forced climate migrants in the future, it also true that already existing complementary protection systems—be they of universal or regional scope—may actually be fit to cover this category of individuals. Chapter 11 thus investigates how and if these systems can adapt to ongoing current social changes and cover forced climate migrants under their protection umbrella.

CHAPTER 10

The Role of the UNFCCC and of the Global Compacts for Refugees and Migration

Starting with the 2010 Cancun Agreements, the UNFCCC system began to openly address climate change-related migration, even though in a very soft and cautious manner. As we already saw, the Cancun Agreement invited state parties to undertake "*measures to enhance understanding, coordination and cooperation with regard to climate change induced displacement, migration and planned relocation, where appropriate, at the national, regional and international levels*".[1] At an intergovernmental level, this call for a better understanding of climate-related displacement stimulated the governments of Switzerland and Norway to launch the Nansen Initiative in 2012. Their aim was to gather governments and experts to assess the state of the art concerning disaster-related displacement, state practices dedicated to it, and to fill the legal gap in which the persons involved to find themselves. The Initiative developed an Agenda, published in 2015, for improving the protection of people displaced across borders by natural disasters and slow onset effects of climate change[2] and this helped to build consensus on key principles and

[1] UNFCCC, Cancun Adaptation Framework (2011), Decision 1/CP.16, UN Doc. FCCC/CP/2010/7/Add.1, Article 14, letter f.

[2] Martin (2017). Toward an Extension of Complementary Protection? In Mayer & Crépeau (Eds.), *Research Handbook on Climate Change, Migration and the Law*. Cheltenham: Elgar, p. 454.

G. Sciaccaluga, *International Law and the Protection of "Climate Refugees"*,
https://doi.org/10.1007/978-3-030-52402-9_10

elements on how to protect individuals displaced across borders in the context of natural disasters.

The UNFCCC's willingness to address climate-related displacement also shined through from the 2015 Paris Agreement, which called for the creation of a task force to study the issue in depth. In September 2018, this "Task Force on Displacement" released its final report, which offers a comprehensive study on the challenges and opportunities to minimize and address climate-related displacement. In the same year, the UN General Assembly adopted the Global Compact on Refugees (GCR) and the Global Compact for Migration (GCM), two groundbreaking soft law tools that aim at enhancing international cooperation worldwide in the field of migration and international protection.

Altogether, these tools can be considered as the "embryo" of what might become in the near future a legal system dedicated to the protection of forced climate migrants. This chapter critically examines these trends, addressing how they may fit in a systemic evolution of international law to provide responses to the complex phenomenon of "climate refugees".

Developments Under the UNFCCC

As we saw, the UNFCCC addressed the issue of climate migration for the first time in 2010 with the Cancun Agreements. When we consider that the UNFCCC was adopted in 1992 and that the 1990 IPCC report already noted the climate-induced migrations would represent an extremely pressing international challenge, we can appreciate that it took quite a lot for the matter to be openly discussed under the treaty's auspices. This is no surprise however when we think through the evolution of this particular legal system as a whole. In fact, all scientific uncertainties (and consequently all political and legal hesitations) that surround(ed) climate change have considerably slowed down any progress in the years. And this possibly even more so when it came to discussing the politically sensitive topic of (inter)national migrations deriving from global warming. Notwithstanding these inevitable slowdowns, it is sound to include climate displacement under the UNFCCC once we acknowledge that the kind of human mobility we are dealing here with is effectively created or exacerbated by anthropogenic global warming.[3] A gradual recognition and inclusion of the topic within the UNFCCC is probably the fittest way to encourage the creation of protection mechanisms for the category of forced climate migrants, since this would avoid the necessity to create ad hoc universal frameworks, whose political feasibility appears far from our grasp.

The celebrated 2015 UNFCCC's Conference of Parties (COP 21) held in Paris saw a further evolution of the trend commenced in Cancun. Even though the topic did not enter the Paris Agreement itself, it was specifically addressed in its attached and non-binding Decision. In the Loss and Damage section, the Executive Committee of the Warsaw International Mechanism was asked to create a task force dedicated to climate displacement, with the objective of "*develop[ing] recommendations for integrated approaches to avert, minimize and address displacement related to adverse impacts of climate change*".[4] Experts on human mobility and representatives of both the Global South and the Global North compose the Task Force, which includes members of the International Labour Organization (ILO), the UNEP, the International Federation of Red Cross and Red Crescent Societies (IFRC), the UNHCR, the IOM, and of the Platform on Disaster Displacement. One year later, COP 22 in Marrakesh encouraged again parties to act in relation to climate migration, also through bilateral and multilateral initiatives.[5] Indeed, this evolution witnesses to a certain extent the parties' willingness to cope with our topic, with the UNFCCC that is gradually taking the responsibility for those who migrate due to climate change.

The Task Force on Displacement (TFD) released its final report in 2018, before COP 24 in Katowice, Poland. It explicitly recognizes that "slow onset events may contribute to decreased ecosystem services [...] that are vital for human survival [and that] the scarcity of vital resources

[3] Atapattu (2018). A New Category of Refugees? "Climate Refugees" and Gaping Hole in International Law. In Behrman & Kent (Eds.), *"Climate Refugees" Beyond the Legal Impasse?* New York: Routledge, p. 48.

[4] Decision 1/CP.21, Adoption of the Paris Agreement, UN Doc. FCCC/CP/2015/10/Add.1, para. 49.

[5] UNFCCC COP 22, Marrakech (2017), Report of the Conference of the Parties on its twenty-second session, Decision 3/CP.22, Warsaw International Mechanism for Loss and Damage associated with Climate Change Impacts, para. 9.

may in turn lead to a serious disruption of livelihoods[6]". Quite obviously then, it states that the scale and nature of resulting human displacement and mobility "will largely depend on the adaptation, disaster risk reduction and development policies that are implemented not only to mitigate the impacts of slow onset events, but also to facilitate, initiate and/or manage migration as a positive strategy and planned relocation as a last resort option".[7] The TFD's work hence stands alongside what we pointed out in Chapter 2 with regard to the role of ecosystem services and their crucial importance to human conditions, and shows that the UNFCCC is now aligned with the scientific knowledge at disposal today. In this context, it is still too soon to establish clear and direct legal provisions under the system: we are still at an embryonic stage. Interestingly though, UNFCCC's parties in Katowice welcomed and endorsed the TFD's recommendations and decided to extend its mandate for another five years.[8]

Broadly, the recommendations endorsed in Poland strongly focus on the need to enhance international cooperation for a better understanding of the phenomenon, thereby providing rather generic indications. However, there are some specific points that may in the long run represent starting points for the development of protection measures dedicated to forced climate migrants. For instance, parties are invited to "*formulat[e] laws, policies and strategies [...] to avert, minimize and address displacement related to the adverse impacts of climate change [...], taking into consideration their respective human rights obligations and, as appropriate, other relevant international standards and legal considerations*".[9] They shall, in addition, continue to share practices in "*providing assistance to, and protection of, within existing national laws and international protocols and conventions, as applicable, affected individuals and*

[6] Report of the Task Force on Displacement (TFD Report), Advanced unedited version of 17 September 2018. https://unfccc.int/sites/default/files/resource/2018_TFD_report_17_Sep.pdf. Accessed 14 March 2020, para. 61, letter d.

[7] Ibid., para. 61, letters b and d.

[8] UNFCCC COP 24, Katowice (2018), Report of the Executive Committee of the Warsaw International Mechanism for Loss and Damage associated with Climate Change Impacts, Decision -/CP.24.

[9] Ibid., Annex "Recommendations from the report of the Executive Committee of the Warsaw International Mechanism for Loss and Damage associated with Climate Change Impacts on integrated approaches to averting, minimizing and addressing displacement related to the adverse impacts of climate change", para. 1, letters g and i.

communities", and "*apply international legal instruments and normative frameworks, as appropriate*".[10]

Noteworthy is that these points confirm that states are asked to address the issue of "climate refugees" by taking into account their human rights and international environmental law obligations. Hopefully, the thesis proposed in this book might prove useful in demonstrating why and how these obligations should effectively apply in relation to the protection of forced climate migrants. In fact, in Part II we suggested that these individuals should be provided with international protection abroad when they cannot enjoy basic human rights in their country of origin because this has become (or is on the verge of becoming) uninhabitable because of environmental degradations that are caused by climate change. States are under the obligation to protect non-derogable fundamental rights universally and, as we pointed out in Chapters 6 and 7, international law is evolving toward an ever-stricter obligation to adopt measures to mitigate global warming and to adapt to its consequences. When read together with the recommendations endorsed by COP 24 in Katowice, these obligations stemming from human rights and climate change international law prove, from a theoretical legal standpoint, the necessity to develop policies and measures relating to forced climate migrants under the UNFCCC's auspices. In a medium-term perspective, the trend commenced in 2010 with the Cancun Agreements can be held as a significant attempt by the UNFCCC's to become the comprehensive international *forum* to address climate-related displacement. The extension of the TFD's mandate for a further five years, with the laying down of a detailed action plan for 2019–2021 go in this direction. Of course, there is still a long way to go, but in the last ten years, the UNFCCC's system has gradually evolved up to the point of recommending its parties to formulate law and policies to address climate-related displacement by taking into account their human rights obligations. And in this context it sobers to recall what the UN Human Rights Committee established in January 2020 in occasion of the *Teitiota case*, namely that states should not reject (*refouler*) individuals who flee from climate change effects when their fundamental rights are in peril.[11]

[10] Ibid., para. 1, letters h (iii) and d.

[11] Please refer to Chapter 8 for more information.

If this general trend continues, it might be possible in the next years to witness further developments, with the UNFCCC that, in full respect of its name, becomes a "framework" under which more and more climate change-related challenges will be dealt with. By starting only with mitigation concerns in its first years, the Convention has in fact progressively passed to addressing both adaptation and loss and damage issues as well. Its recent incorporation of human rights law witnesses that the system is becoming more open to other branches of international law. The composition of the TFD, which is mainly of human mobility specialists rather than of climate/environmental experts, also witness this evolution. If there is a legal framework where to address climate migration, and hence where to establish guidelines, recommendations, or eventually even binding provisions for the protection of "climate refugees", that legal framework is the UNFCCC. To reach a similar objective, this book suggests that it is sound to carefully circumscribe the category of those who should effectively be protected in response to their forced migration. Part I and II served this purpose. We defined who forced climate migrants are and outlined the legal reasoning by virtue of which they shall be protected under international law so as to provide a theoretical legal argument that should support the development of future protection mechanisms. The assumption here is that states are and will be willing to enlarge their protection duties if and only if these are to be granted to a limited and manageable number of people. Forced climate migrants, as understood in the present book, may encounter the states' favor in this sense.

The Role of the Global Compacts for Refugees and Migration

In 2018, Traore Chazalnoel and Ionesco argued that whereas the inclusion of migration within the work of the UNFCCC could be "well documented", with "the topic [...] firmly anchored within the negotiations on international climate policy", the same level of acknowledgment could not be appreciated within global migration policy.[12] They saw in the then-ongoing negotiations on the two Global Compacts for

[12] Traore Chazalnoel & Ionesco (2018), *Advancing the Global Governance of Climate Migration Through the United Nations Framework Convention on Climate Change and the Global Compact on Migration. Perspectives from the International Organization for*

Refugees and Migration a promising opportunity to start considering environmental and climate factors in human mobility frameworks as well.

By the end of that year, the UN General Assembly adopted the two groundbreaking instruments with the wide-ranging objective of enhancing international cooperation in the fields of international protection and migration management. The Global Compacts for Refugees and Migration represent a milestone. They come as a result of the momentous adoption of the 2015 Sustainable Development Goals, the programmatic political document that sets down the international community's development objectives for 2030. The two Compacts are non-binding, cooperative frameworks that build on the commitments agreed upon in 2016 by UN member states in the New York Declaration for Refugees and Migrants, which clearly acknowledged the interrelation between climate change and migration.[13] While the two Compacts mainly aim at facilitating international cooperation in the field of international protection and migration, they also present some remarkable points that witness the embryonic evolution we have been referring to so far and that is currently going on regarding the topic of climate change and human mobility.

The Global Compact on Migration (GCM) starts from the assumption that no state alone can address migration issues: the challenges and the opportunities linked to migration flows in the twenty-first century are far too big to be addressed by any state acting alone, and this shows again, as we highlighted in Chapter 2, that the international system is moving toward the gradual affirmation of interdependence as a paramount concept rather than predominantly relying on the outmoded notion of national interest. With a similar trend, there is a need to enhance international cooperation among all relevant actors and encourage the establishment of bilateral or multilateral migration policies to facilitate legal migration pathways and to prevent illegal ones. With its overarching approach, which addresses both the causes and consequences of migration worldwide, the GCM invites states to mitigate the "*adverse drivers and structural factors that hinder people from building and maintaining sustainable livelihoods in their countries of origin, and so compel*

Migration. In Behrman & Kent (Eds.) (2018). *"Climate Refugees" Beyond the Legal Impasse?* New York: Routledge, pp. 103 and 104.

[13] UN General Assembly Resolution (2016). New York Declaration for Refugees and Migrants, UN Doc. A/RES/71/1, para. 43.

them to seek a future elsewhere".[14] For our purposes, the Compact interestingly and explicitly recognizes that environmental changes and global warming are important drivers of migration. Paragraph 18 dedicates an entire section to "*natural disasters, the adverse effects of climate change, and environmental degradation*".[15] Fundamentally, it urges states to favor political, economic, social, and environmental conditions that allow people to lead decent and sustainable lives in their countries of origin. With a view to environmental degradation only, states are recommended to ensure that ecosystem changes do not compel individuals and communities to seek decent living conditions abroad through irregular migration.

Thus, the GCM unequivocally includes climate change as a decisive migration push-factor, and, in line with what the Task Force on Displacement recommends, it calls for international efforts to better understand, predict, and address human movements resulting both from sudden and slow-onset natural disasters, environmental degradation, and climate change.[16] Hence, the approach followed here by the GCM is comparable to the recent UNFCCC's one: states still essentially aim at better understanding the topic as a whole with a view to properly addressing it in future, when more certainties will be at their disposal. Their current purpose is to foster future international policies having the twofold objective of mitigating environment and climate-related migration and protecting the human rights of those who are and will be more vulnerable to it.

With a view also to forced migrations, the GCM calls for "*solutions for migrants compelled to leave their countries of origin owing to slow-onset natural disasters, the adverse effects of climate change, and environmental degradation, such as desertification, land degradation, drought and sea level rise, including by devising planned relocation and visa options, in cases where adaptation in or return to their country of origin is not possible*".[17] This paragraph directly refers to forced climate migrants and shows that the GCM was adopted with the clear awareness that *non refoulement*

[14] Global Compact for Safe, Orderly and Regular Migration (GCM), UN Doc. A/RES/73/195, Preamble, para 11.

[15] Ibid., para. 18 letters h - l.

[16] Ibid., para. 18, letter h.

[17] Ibid., para. 21, letter h.

should apply in cases of climate-related migration when returning to the country of origin is not possible. Paragraph 37 of the document is also crucial in this regard. When addressing the commitment to facilitate safe and dignified return procedures to their origin countries for all international migrants, it substantiates the prohibition to return individuals to situations of irreparable harm, in accordance with human rights obligations. By so doing, the GCM implicitly upholds that return to a region or a country that has become irreparably uninhabitable due to ecosystemic failures induced by climate change is to be excluded under international human rights law, something that has for instance also been confirmed by the UN Human Rights Committee in the occasion of *Teitiota* in January 2020.[18]

The GCM then strongly focuses on the necessity to implement regular migration pathways.[19] The UNFCCC's Task Force on Displacement as well called for facilitating "orderly, safe, regular and responsible migration and mobility of people [...] in the context of climate change by considering the needs of migrants and displaced persons, communities of origin, transit and destination, and by enhancing opportunities for regular migration pathways [...]".[20] Intergovernmental migration policies may indeed have a significant impact on the management of climate-related migration, especially in some regions. When politically feasible, gradual, and agreed upon migration corridors may solve some of the problems arising alongside the climate migration challenge. International migration policies following for example the Pacific Access Category's model can allow a gradual settlement of the involved communities in the receiving states.[21] Of course, as we already pointed out, similar initiatives might prove effective only in particular areas: whereas the planned and gradual migration of some hundreds of thousands of inhabitants of "drowning" Pacific atolls onto the territory of some (developed) neighboring states can probably be implemented in a safe, dignified, and manageable manner, the same cannot be assured when it comes to far larger numbers of people who (should) migrate to densely populated and less developed neighboring

[18] Please refer to Chapter 8.

[19] GCM, Objective 5.

[20] TFD Report (op. cit.), para. 1, letter g (vi).

[21] Please refer to Chapter 4.

states (such as it might be the case in the Ganges Delta region or in some parts of Africa for instance).

IOM's Regional Consultative Processes (RCPs) clearly represent a sound context where to discuss these options at a (sub)regional level. RCPs are state-led channels of communication on migration issues between countries of a common geographical area and with potentially coinciding interests. They can operate as informal bodies or can be affiliated to regional institutions and organizations. Notwithstanding this lack of uniformity, their common purpose is to gather state representatives, international organizations, and occasionally non-governmental organizations too to exchange information on migration-related issues within their geographical area of interest. The Platform on Disaster Displacement for instance suggests that the GCM's outcome also derives from the work and progress made under RCPs' work worldwide.[22] Indeed their informal and elastic structure might also prove useful in the following years with regard to the development of regional migration policies (in)directly addressing climate-related migration problems.

As a whole, the fact that the GCM comprehensively acknowledges the role of climate change and environmental degradation as migration drivers is surely noteworthy with a view to enhancing the adoption of dedicated measures. As the Task Force on Displacement put it in 2018, "this historical migration policy achievement could trigger a review of existing national human mobility policy frameworks in line with the GCM provisions. New mobility policy frameworks could also be developed on the basis of this Compact, opening the possibility to further mainstream climate and environmental dimensions".[23] The explicit reference in the GCM to natural disasters, environmental degradation, and climate change as migration push-factors is surely important in a world in which some (very few) governments even try to deny the existence of global warming *tout court*. On the other hand, however, the Compact fundamentally recommends states to study the issue further so as to adopt responses in the future: given that climate change produces and increases international migrations, cooperative policies must be developed to better understand what the phenomenon as a whole is about, minimize the problems at their

[22] Platform on Disaster Displacement (2018). In Behrman & Kent (Eds.), *"Climate Refugees" Beyond the Legal Impasse?* New York: Routledge, pp. 139 and 140.

[23] TFD Report (op. cit.), para. 65.

source and, where necessary, encourage regular and legal migration, both nationally and regionally. We can conclude, hence, that the GCM's impact is mainly symbolic and exhortative, but this is, truly speaking, in line with the nature of every universal, all-encompassing soft law instrument.

The Global Compact on Refugees deals again with the "protection side" of our phenomenon. It recognizes that there is an interaction between climate change, environmental and natural disasters, and refugee movements: "*while not in themselves causes of refugee movements, climate, environmental degradation and natural disasters increasingly interact with the drivers of refugee movements*".[24] The UNHCR has interestingly stressed here the importance of choosing the wording "*increasingly interact*", because it highlights the awareness of the growing importance of the interplay between environmental pressures and forced migration. With its comprehensive approach, the GCR refers to the need for reducing disaster risks and includes refugees in disaster risk reduction strategies. With a view to the potential future development of dedicated protection and assistance measures, it recalls existing international obligations so as to "*avoid protection gaps and enable all those in need of international protection to find and enjoy it*",[25] and it also calls for all "*stakeholders [...] [to] provide guidance and support for measures to address [...] protection and humanitarian challenges. This could include measures to assist those forcibly displaced by natural disasters, taking into account national law and regional instruments as applicable, as well as practices such as temporary protection and humanitarian stay arrangements*".[26] These points mirror the awareness concerning the legal gaps in which forced climate migrants find themselves and request states to cope with these as soon as possible. As a whole, The GCR is thus trying to incentivize, just as the UNFCCC and the GCM do, the development of protection mechanisms for those who are compelled to leave their countries because of climate change-related disasters.

Taken together, the adoption of the two Compacts comes as interesting and perhaps even historical news in the view of addressing the challenge of "climate refugees". The GCM opens both the way for the

[24] Global Compact on Refugees (GCR), General Assembly Official Records Seventy-Third Session Supplement No. 12 (A/73/12 [Part II]), para. 8.

[25] Ibid., para. 61.

[26] Ibid., para 63.

adoption of cooperative international migration policies and of protection measures for forced climate migrants, explicitly and exhaustively referring to climate change and environmental degradation as decisive migration push-factors for the years to come, something that may eventually induce states and international organizations to develop dedicated responses in the next future. The GCR also recognizes the interplay between environmental/climate pressures and refugee movements. It furthermore asks states to fill existing international protection gaps and to provide responses for those who are in need for international protection but currently not covered by any normative framework.

As a whole, the two Compacts clearly suggest therefore that the international legal system is in an embryonic but promising stage of development toward the creation of a multifaceted and inclusive system dedicated to the management of climate change-related migrations.

CHAPTER 11

The Role of Complementary Protection

As known to any scholar in our field, research has clearly shown over the years that under current international law the category of forced climate migrants is not adequately covered. Refugee law, as presently structured and applied, is not suitable to guarantee protection to these people. To do so, explicit modifications of the text of the 1951 Refugee Convention are needed,[1] since its definitional approach, clearly reflecting a post-Holocaust regime, does not leave room for "environmental" or "climate refugees". The issue of extensively interpreting the 1951 Convention is nothing new. In the Final Act of the Conference that adopted this treaty, we may for instance read that there was "*the hope that the Convention [...] will have value as an example exceeding its contractual scope and that*

[1] In this regard, see as a matter of example Cooper. With the aim of adapting the 1951 Geneva Convention on Refugees to the spirit of the Universal Declaration of Human Rights, in particular where it establishes that "Everyone has the right to a standard of living adequate for the health and well-being of himself and of his family [...]" (Article 25), Cooper proposes a new definition of "refugee" which covers so-called environmental refugees too: "[...] anyone in justified fear of being persecuted for his race, his religion, his citizenship, his membership of a particular social group or his political opinions [...] "or victim" of degraded environmental conditions that threaten his life, his health, his livelihoods or the use of natural resources". Cooper (1998). Environmental Refugees: Meeting the Requirements of the Refugee Definition. *New York University Environmental Law Journal*, 6.

G. Sciaccaluga, *International Law and the Protection of "Climate Refugees"*,
https://doi.org/10.1007/978-3-030-52402-9_11

all nations will be guided by it in granting so far as possible to persons in their territory as refugees and who would not be covered by the terms of the Convention, the treatment for which it provides" (para. IV, recomm. E). Some authors have claimed that this is the inspirational basis for the development of complementary protection systems and that it can in some way also be used to justify an extensive interpretation of the Convention itself.[2] However, this last position does not appear entirely acceptable in light of Articles 31–33 of the 1969 Vienna Convention on the law of treaties. In fact, the Final Act of the Conference should not be considered as an instrument allowing an objective interpretation of the treaty, but merely as an auxiliary tool. The 1951 Refugee Convention leaves in its binding part little room for extensive interpretations when it comes to defining refugees. Thus, the recommendation of the Final Act merely called (and is *still* calling) for states to create *ex novo* complementary protection instruments, and this is precisely what happened in the last seven decades.

In a medium-term perspective—that is when the phenomenon of "climate refugees" will in all probability become an issue of significant humanitarian concern worldwide—more room for granting protection rights to forced climate migrants will be found in complementary protection systems. Subsidiary or complementary protection systems consist in the set of rules that originate from humanitarian and human rights law and that apply the *non refoulement* principle in addition to the hypotheses envisaged by the 1951 Refugee Convention. More precisely, complementary protection applies when an asylum seeker runs the risk, in the case of return to the country of origin, of incurring in inhuman, degrading or otherwise harmful treatments to human dignity. Although not explicitly recognized in specifically dedicated international instruments, the existence of a protection regime that is complementary to the 1951 Refugee Convention can be inductively demonstrated through various international, regional, and national provisions regulating the protection of

[2] Spijkerboer (2002). Subsidiarity in Asylum Law: The Personal Scope of International Protection. In Bouteillet-Pacquet (Ed.), *Subsidiary Protection of Refugees in the European Union: Complementing the Geneva Convention?* Brussels: Bruylant, pp. 19 ff.

human rights.[3] Ultimately, these provisions altogether constitute a significant normative basis for complementary protection under international law.

For the purposes of this book, complementary protection plays a significant role, because in cases of migrations caused by climate change, the *non refoulement* principle could be activated and applied, thereby granting some form of international protection to forced climate migrants. An individual fleeing a devastated environmental context could enjoy a right of permanence and protection in a foreign state if his/her repatriation to the country of origin entails the risk of inflicting inhuman or degrading treatments, capable of depriving the person of fundamental and inalienable rights. Generally, inhuman or degrading treatments are made of serious deprivations of primary rights, such as the rights to life and human dignity. These deprivations can arise when a person faces the risk of death or torture in his/her country of origin and, interestingly with a view to environmental degradations, such deprivations can also arise when the person would be obliged to live in exceptionally dramatic socio-economic conditions.

The 1951 Refugee Convention represents the "cornerstone" and the original foundation of the *non refoulement* principle, but under complementary protection systems, *non refoulement* enjoys however a wider application. A state may be obliged to apply complementary protection frameworks either arising from national or international human rights provisions. It has then no right to reject (*refouler*) an individual under its jurisdiction where repatriation involves the risk, for that particular person, to suffer from inhuman or degrading treatment. In a limitation of its sovereign powers dictated by human rights protection concerns, the host country is denied the possibility of removing the subject from its territory: if it did so, it would contribute, albeit indirectly, to violations of primary and inalienable human rights.

[3]See McAdam (2011). Climate Change Displacement and International Law: Complementary Protection Standards. UNHCR Legal and Protection Policy Research Series, doc. PPLA/2011/03. https://www.unhcr.org/4dff16e99.pdf. Accessed 10 February 2018. See also McAdam (2007). *Complementary Protection in International Refugee Law*. Oxford: Oxford University Press. Goodwin-Gill (1986). Non-refoulement and the New Asylum Seekers. *Virginia Journal of International Law*, 1986, pp. 897 ff.

Among these, the first and the basis of all others is obviously the right to life, protected, at a universal level, by Article 3 of the Universal Declaration on Human Rights,[4] as well as by Article 6 of the ICCPR,[5] and Article 6 of the CRC.[6] As already highlighted, the right to life also finds its specifically dedicated regulation at a regional level.[7] The right to life stands at the very core of the exercise of every other existing subjective right. It protects a "good" that must be guaranteed in an absolute and binding way and it is essentially for this reason that a risk to lose one's own life entails the triggering of the *non refoulement* obligation every time a person's repatriation could lead to death.[8] By its own nature, the right to life is intimately connected to the enjoyment of every other fundamental right. In our study, climate change, through effects such as the reduction of cultivable soils can contribute—especially in particularly vulnerable geographical areas—to the depletion of livelihoods, thus directly and indirectly affecting the possibility of exercising the right to life.

Non refoulement applies also when there is a risk of incurring inhuman or degrading treatment. In the context of universal human rights treaties, the legal sources for such eventualities are to be found in Article 7 of the ICCPR[9] and in Article 3 of the Convention against Torture and Other

[4] Universal Declaration of Human Rights, adopted on 10 December 1948, Article 3 "*Everyone has the right to life, liberty and the security of person*".

[5] International Covenant on Civil and Political Rights, adopted 16 December 1966, entered into force 23 March 1976., Article 6: "*Every human being has the inherent right to life. This right shall be protected by law. No one shall be arbitrarily deprived of his life*".

[6] Convention on the Rights of the Child, adopted on 20 November 1989, Article 6: "*States Parties recognize that every child has the inherent right to life*".

[7] Article 2 of the European Convention on Human Rights; Article 2 of the American Convention on Human Rights; Article 4 of the African Charter on Human and Peoples Rights; Article 4 of the Arab Charter on Human Rights.

[8] See in this regard the activity of the UN Human Rights Committee (HRC) (2004). General Comment No. 31: The Nature of the General Legal Obligation Imposed on States Parties to the Covenant, para. 12. https://www.refworld.org/docid/478b26ae2.html. Accessed 20 March 2020. See also UN Committee on the Rights of the Child (CRC) (2005). General Comment No. 6 (2005): Treatment of Unaccompanied and Separated Children Outside Their Country of Origin, para. 27. https://www.refworld.org/docid/42dd174b4.html. Accessed 12 February 2020.

[9] ICCPR, Article 7: "*No one shall be subjected to torture or to cruel, inhuman or degrading treatment or punishment. In particular, no one shall be subjected without his free consent to medical or scientific experimentation*".

Cruel, Inhuman Punishments or Degrading Treatments.[10] According to Article 7 of the ICCPR for instance, "*no one shall be subjected to torture or to cruel, inhuman or degrading treatment or punishment*". Through these provisions, state parties have subjected themselves to a fairly incisive human rights protection system (at least on paper), under which the expulsion of an individual is prohibited if he/she runs the risk of being humiliated, tortured, or "dehumanized" in case of return to the origin country.

From a conceptual point of view, one can (and probably even should) consider the set of rules capable of activating the *non refoulement* principle under complementary protection systems through a holistic perspective. In this perspective, as the UN Human Rights Commission pointed out in 2005, "the right to life encompasses existence in human dignity with the minimum necessities of life".[11] Understood as such, the right to life can only be guaranteed if the right to health, the prohibition of torture, or degrading treatment and, more generally, minimally acceptable standards of living are also respected.

In Chapter 8, we saw that extremely unfavorable environmental conditions have the potential to determine situations where the effective exercise of these rights cannot be guaranteed. Along with this, we saw how global warming, through its negative impacts, can in some cases be considered as the prevailing factor of certain environmental degradations that make it impossible to effectively enjoy such rights. In this sense, therefore, complementary protection could be invoked to protect forced climate migrants: when the territory of a state becomes uninhabitable (or habitable only in inhuman and degrading conditions) due to anthropogenic climate change, an individual fleeing it should be provided with international protection in the receiving state. However, in order for this protection to be granted, the risk faced by the asylum seeker must be currently in place or, at least, imminent. This is a requirement

[10] Convention against Torture and Other Cruel, Inhuman or Degrading Treatment or Punishment, adopted 10 December 1984, entered into force 26 June 1987, Article 3: "*No State Party shall expel, return (*"refouler"*) or extradite a person to another State where there are substantial grounds for believing that he would be in danger of being subjected to torture*".

[11] UN Commission on Human Rights (2015). Human Rights Resolution 2005/16: Human Rights and Extreme Poverty (E/CN.4/RES/2005/16), para. 1b. https://www.refworld.org/docid/45377c340.html. Accessed 10 March 2020.

that makes it difficult to obtain international protection when an individual is "fleeing" his or her own country due to progressive and gradual climate/environmental degradations that have not *yet* made the territory completely uninhabitable.

We may here advocate for an extensive application of the precautionary principle, arguing that complementary protection should be provided to forced climate migrants even when their territory is *about to become* uninhabitable. The precautionary principle applies in international environmental law, not in human rights law. However, since the condition of forced climate migrants is strictly connected to and dependent on the environmental condition of their homeland, a case can be made for linking these two branches of law: where there are threats of serious or irreversible environmental damages, states are requested to enact environmental protection policies; it would be sound, therefore, to apply this rationale also to protect forced climate migrants where there is relevant scientific evidence that their territory will not be suitable anymore to guarantee the enjoyment of fundamental human rights. However, if we rely only on current rules and their interpretations, a future risk, even if scientifically based, is not sufficient to activate protection mechanisms in application of complementary protection frameworks. This shows, perhaps, that evolutionary and innovative interpretations of existing instruments are today becoming necessary. In this regard, due attention is to be dedicated to the developments recently occurred both with the adoption of the Global Compact on Migration and coming from the UN Human Rights Committee's decision on *Teitiota*. Both the Compact and the decision in fact indicate that *non refoulement* should also apply *before* the actual manifestation of an environmentally irreparable harm that renders life with dignity unavailable in the country of origin,[12] and they show that international law is at present moving toward innovative and extensive interpretations of the instruments at our disposal today. As for now however, a future risk, although well founded, does not immediately constitute a violation of fundamental human rights. Asylum pleas presented in the absence of a current or imminent threat are likely to fail, as the host state is not required to protect a foreigner where he or she

[12] See the Global Compact on Migration, para. 37, when it states that returning a migrant to his/her origin country should be prohibited in presence of a foreseeable risk of irreparable harm. See also UNHCR (2020). *Teitiota v. New Zealand* (cit.), para. 9.11.

does not run the risk, in case of *refoulement*, of incurring in the short run in inhuman or degrading living conditions.

Put simply, the moment in which an asylum plea is submitted is a determining factor. Current and existing case law does not foresee, in fact, that preventive movements driven by the expectation (albeit scientifically based) of the worsening of living conditions due to climate or environmental pressures are able to activate the principle of *non refoulement*. A similar interpretation should and could probably change in the near future, especially with regard to the most vulnerable individuals that come from insular micro-states at risk of submersion, or where there is already a fair degree of certainty regarding the progressive worsening of environmental conditions and socio-economic factors due to climate change.

The opinion of the UN Human Rights Committee on the link between the threat to the right to life (as regulated by the ICCPR) and the potential use of nuclear weapons is also well suited to determine whether an appeal for a violation of the right to life could be accepted in presence of risks related to gradual climate or environmental degradation. The Committee declared that, to consider someone a victim of a deprivation of the right to life for the conduct of a state, "he or she must show either that an act or an omission of a state party has already adversely affected his or her enjoyment of such right, or that such an effect is imminent".[13] The analogy between fear of damage caused by the use of nuclear weapons and climate change has its *raison d'être*, given the lack of "predictability" and high destructiveness of both. McAdam as well insists in this sense: "one could imagine a similar rationale being used in the climate change context, given uncertain timescales about when climate change impacts will be most severe, and the 'constantly evolving situation' with respect to human adaptation and resilience".[14]

With a view complementary protection on a regional scale, a reference certainly deserves to be made to the European Union, which, in its special legislation on asylum, provides for protection schemes for a much

[13] UN Human Rights Committee (2006). *Aalbersberg* et al. *v. Netherlands*. Communication No. 1440/2005, UN Doc. CCPR/C/87/D/1440/2005, para. 6.3. http://hrlibrary.umn.edu/undocs/1440-2005.html. Accessed 10 January 2020.

[14] McAdam (2012). *Climate Change, Forced Migration, and International Law*. Oxford: Oxford University Press, p. 86.

wider category of people than that covered by the 1951 Refugee Convention.[15] Before the EU, in reality, among the various international human rights frameworks, one of the most innovative is the Council of Europe's ECHR: its case law is an important reference point for understanding the problems linked to the possible application of the principle of *non refoulement* to forced climate migrants. As known, the Court of Strasbourg can "update" the ECHR's contents in the light of current historical and social developments. Usually, in the context of the ECHR, the activation of the *non refoulement* principle occurs in presence of violations of Articles 2 and 3. They guarantee, respectively, the right to life and the prohibition of torture and inhuman or degrading treatment. The two provisions act very often together, precisely because a holistic reading of the right to life presupposes that, for this to be respected, an individual cannot be subjected to inhuman or degrading treatment that may damage his or her intrinsic dignity as a human being. No *non refoulement* request has in fact ever been accepted only on the basis of Article 2,[16] but always in conjunction with Article 3, and this is because where a violation of the latter is ascertained, the need to consider possible violations to the right to life almost automatically falls, becoming unnecessary for the purposes of attributing international protection rights.

Consequently, to understand what role a system such as the ECHR can play in potential cases of *non refoulement* related to migrations forced by climate change, it is necessary to analyze how Article 3 is applied in relation to comparable situations. The absolute and mandatory scope of the prohibition of torture or inhuman/degrading treatment was established for the first time in 1989, on the occasion of *Söring v. United*

[15] See Directive 2011/95/UE (Qualification Directive), Art. 15. Pursuant to this provision, international protection is to be granted to individuals who, in case of *refoulement*, face the risk of being sentenced to death, torture or inhuman or degrading treatment, or that face serious life-threat resulting from indiscriminate violence in situations of internal or international armed conflicts. Refer also to Directive 2013/33/UE; Directive 2013/32/UE; and Directive 2001/55/CE. On the EU asylum system, see *ex multis* Celle (2011). The Right of Asylum: Toward a Common European System. Impact on the Mediterranean Area. In Bottaro Palumbo & Danisi (Eds.), *Civil Rights Protection and the Rights of Migrants in the Framework of the Mediterranean Cooperation*. Genoa: Genova University Press. See also Munari (2016). The Perfect Storm on EU Asylum Law: The Need to Rethink the Dublin Regime. *Diritti Umani e Diritto Internazionale*, pp. 517 ff.

[16] An ECHR Article 2 violation in a *refoulement* case has been ascertained in only one case, but always in conjunction with Article 3: see European Court of Human Rights, *Bader and Kanbor v. Sweden*, Application No. 13284/04, 8 November 2005.

Kingdom,[17] in which the extradition to the United States of an individual convicted of murder in the United States, but placed under British jurisdiction at the time of the appeal, would have put his life at risk because of the detrimental impacts of the death row phenomenon and of the existence, in the United States, of the death penalty. The Court of Strasbourg established then—and later always reiterated this position—that the Convention prohibits any type of derogation or exception to the application of Article 3, which is absolute: state parties cannot in any way allow an individual under their jurisdiction to reach a country where his or her non-derogable rights are seriously endangered. The absolute non-derogability of this disposition entails that it cannot be balanced anyhow by the interests of the single state, something that is on the contrary allowed when socio-economic rights come at stake.

The Court of Strasbourg has often underlined that an inhuman treatment, to be such, must reach a minimum level of severity and include a concrete bodily injury or intense suffering of physical or mental kind.[18] In a complementary way, a treatment becomes degrading when it "humiliates or degrades an individual, showing a lack of respect for - or diminishing - his human dignity, or when it creates a feeling of fear, anguish or inferiority capable of breaking the moral or physical resistance of the individual".[19] In all this, the minimum level of severity "depends on all the circumstances of the case, such as the duration of treatment, its mental and physical effects and, in some cases, the sex, age and health status of the victim".[20] An important element with reference to climate change is that the treatment, in order to configure a violation, does not need to be deliberate: a lack of intent to humiliate or degrade does not automatically lead to an exclusion of the applicability of Article 3.[21]

[17] European Court of Human Rights, *Söring v. United Kingdom*, Application No. 14038/88, 7 July 1989.

[18] European Court of Human Rights, *Pretty v. United Kingdom*, Application No. 2346/02, 29 April 2002, para. 52.

[19] Ibid.

[20] European Court of Human Rights, *N v. United Kingdom*, Application No. 26565/06, 27 May 2008, para. 29. See also *Paposhvili v. Belgium*, Application No. 41738/10, 13 December 2016, para. 183.

[21] European Court of Human Rights, *Labita v. Italia*, Application No. 26772/85, 6 April 2000, para. 120.

Now, what is relevant in relation to the category of migrants we are dealing with in this book is that the rationale of Article 3 is also applicable to cases concerning violations of socio-economic rights. By their very nature, the legal frameworks envisaged for protecting human rights tend to focus more on civil and political rights, relegating socio-economic rights to the background. And it is mainly to overcome this setting that a "holistic" or "integrated" approach has been developed in the years by the case law: with this approach, social and economic rights become integral elements of civil and political ones. If there are serious deprivations of the former, also the enjoyment of the latter is endangered. Thanks to this approach, it becomes possible to apply Article 3 and the *non refoulement* principle in case of violations of socio-economic rights, which are relevant in the context of global warming.

At this point, the important question is: can the existence of conditions of socio-economic poverty exacerbated by climate change in the country of origin of an individual integrate a possible deprivation of Article 3 of the Convention and, therefore, activate the *non refoulement* principle in case of asylum pleas?

The threshold beyond which a treatment becomes inhuman or degrading is, as we saw, definitely high: it implies the existence of serious violations of primary human rights. When one enters the field of socio-economic deprivations, this threshold goes even further, to a level where it is difficult to provide effective protection for those who migrate because of climate change (at least in the vast majority of cases). To reach the threshold needed to apply Article 3, the Court Strasbourg has indeed established that a significant decline in the applicant's living conditions—including his or her life expectancy—is not enough. The application of Article 3 in cases of socio-economic deprivation is "justified only when the humanitarian component of the case is so urgent that it cannot be ignored by the authorities of a civilized state. […]. A case related to Article 3 of this kind must [therefore] be based on facts that are not only exceptional, but also extreme".[22] Thus, to activate Article 3, there must be (*i*) a situation of extreme poverty characterized, (*ii*) in the specific individual case, by exceptional circumstances. In itself, extreme poverty is not enough to apply *non refoulement*. When it comes down to it, a state party to the ECHR is not under a general obligation to assist all the needy.

[22] England and Wales Court of Appeal, *N v. Secretary of State for the Home Department*, Case No. C4/2004/0669, 5 August 2004.Civ. 1369, 2003, paras. 38–40.

This obligation is only requested if and when the asylum seeker is in such a critical situation—due to exceptional circumstances—that his/her return to the country of origin, as a consequence of a *refoulement* by the receiving country, would deny him/her access to livelihoods or the most fundamental necessities to life. It remains clear in a similar context that every single case then largely depends on the characteristics of the applicant, such as age, gender, psychophysical conditions, etc. A similar case dates for example back to 2011, when the Belgian government was convicted of violating Article 3 for rejecting an asylum seeker to Greece, exposing him—knowingly—to conditions of detention and life that were considered degrading, consisting "in the disability to provide for their most fundamental needs, such as food, hygiene and housing, in conjunction with the constant fear of being robbed and the total improbability of an improvement in the situation".[23] A further useful case relates to a *non refoulement* obligation imposed on the United Kingdom to protect some Somali citizens, whose return to a Somali refugee camps would have subjected them to such disastrous conditions to be inhuman or degrading. In the judgment, the most interesting aspect consists in the emphasis placed on the conjunction between the presence of a deliberate action of the Somali public authorities and the environmental difficulty characterizing the case: "it is clear that, in this case, despite the drought has contributed to the humanitarian crisis, this is predominantly due to the direct or indirect actions of the parties in conflict".[24] By analogy, therefore, if this interpretation were to be maintained, a violation of Article 3 referable to environmental degradation caused by climate change should be more easily determinable if a deliberate action or inaction by one or more public authorities could be ascertainable. Consequently, it is necessary to ask whether this approach can be applied even in cases where climate change, even if not directly attributable to the state of origin of an asylum seeker, would require, at home, a "positive obligation" of adaptation through policies and regulations aimed at protecting the rights of its citizens in the face of serious environmental degradation caused by climate change.

[23] European Court of Human Rights, *MSS v. Belgium & Greece*, Application No. 30696/09, 21 January 2011, paras. 252–254.

[24] European Court of Human Rights, *Sufi & Elmi v. United Kingdom*, Applications No. 8319/07 and 11449/07, 28 June 2011, para. 282.

More broadly, in any case, systems such as the ECHR obviously do not impose a ban on expulsion by the contracting state simply because in the country of origin of the asylum seeker the living conditions are relatively worse. Maintaining the opposite would amount to imposing an obligation on parties to act actively as guarantors of human rights for the rest of the world. In other words, and exactly as it happens with regard to the application of refugee law under the 1951 Geneva Convention, under Article 3 ECHR it is not possible to obtain international protection only because of the poverty present in other parts of the globe: a person may be rejected in a country where the general standards of life are lower than those in the country in which protection is sought. Such protection is granted only if the deficient exercise of human rights of the subject at home involves a threat to life or exposes him/her to the risk of inhuman or degrading treatment or punishment.[25] Ultimately, and with a view to international protection, the function of regional human rights protection instruments, for which we have taken the ECHR as a model, is not to provide protection from the socio-economic difficulties that can be found in certain geographical contexts. If this were to happen, and if, that is, state parties had the obligation to correct the socio-economic disparities at the global level by guaranteeing a right of asylum to disadvantaged individuals, the Convention, the Court, and, with them, the states themselves would find on their shoulders an enormous and unbearable obligation.[26]

It is therefore for pragmatic reasons that the ECHR does not guarantee protection from socio-economic deprivations, except for exceptional and extreme cases. With regard to forced climate migrants, the very high threshold currently set by the relevant case law with a view to applying *non refoulement* to social, economic (and environmental, we would say) living conditions involves a major implication: it will take probably decades before the negative effects of climate change, interacting with economic and social variables, lead to an activation of the prohibition of *refoulement* in application of Article 3 of the Convention.

[25] In this sense, see also UNHCR (2003). Guidelines on International Protection: "Internal Flight or Relocation Alternative" Within the Context of Article 1A(2) of the 1951 Refugee Convention and/or 1957 Protocol Relating to the Status of Refugees. UN Doc HCR/GIP/03/04, para. 29. https://www.unhcr.org/publications/legal/3f28d5cd4/guidelines-international-protection-4-internal-flight-relocation-alternative.html. Accessed 13 June 2019.

[26] See in this sense also European Court of Human Rights, *F v. United Kingdom*, Application No. 17341/04, 22 June 2004.

Summarizing, the element that the different courts and international bodies have focused on is the risk of "irreparable harm". The Court of Strasbourg itself, starting from the *Söring* case, for instance, has always hinted that other obligations capable of activating the *non refoulement* principle may be implicit in other provisions of the Convention, in addition, of course, to Articles 2 and 3. Nevertheless, the same Court has marked a line of distinction between fundamental rights that are to be considered as mandatory and inalienable, and the others, that do not automatically activate *non refoulement*. The reason behind this distinction is related to what we previously saw with regard to the ECHR objectives. The Convention, exactly as the other regional human rights treaties, does not oblige a contracting state to be the indirect guarantor of fundamental rights on a global scale; consequently, there is a precise limit sets to the rules capable of activating *non refoulement*, since there is a precise legal (although to a certain extent ontologically blurred) distinction between those who have the right to protection and stay in a third state and those who do not.

Finally, we must consider whether the ICESCR may impose complementary protection obligations too. Mainstream opinion in this regard tends to be negative. It is believed, in fact, that the provisions of the Covenant are not in themselves sufficient to activate the *non refoulement* principle. However, part of the doctrine insists that socio-economic rights are, instead, a sufficient basis for international protection.[27] Firstly, it underlines that certain articles of the treaty are immediately binding, such as, for example, those relating to equality between men and women (Article 3). Secondly, Article 2 of the treaty imposes two immediate obligations: the duty to ensure that the rights enunciated are fulfilled, and the obligations to allow the indiscriminate exercise of these. This entails, at least in theory, that a state party in which a significant number of individuals are deprived of their basic needs can be held in violation of its treaty obligations. In all this, of course, a deliberate action (or inaction) adopted by a public authority is necessary for a protection to be invoked.

Coming to a conclusion, although there are in theory, both on an international and regional scale, other rights or dispositions, in addition to the "canonical" ones, namely relating to life and the prohibition of inhuman treatments, capable of activating complementary protection, it

[27] Cf. Foster (2007). *International Refugee Law and Socio-Economic Rights: Refugee from Deprivation*. Cambridge: Cambridge University Press.

remains to be seen, in practice, what these are. To date, the case law has only clarified how these elements are implicitly foreseen, without having specifically dealt with cases capable of identifying a clear set of rules that is actually relevant. In this sense, a New Zealand case in which some "climate refugees" have obtained a right of permanence and protection in New Zealand by virtue of humanitarian protection, a mechanism ascribable to complementary or subsidiary protection, is important.[28] The case, dating back to June 2014, saw the New Zealand Immigration and Protection Tribunal[29] grant protection to a family from Tuvalu affected, among other things, by climate change. The judgment has fed, above all in the media, the idea that "the era of climate refugees" had begun. However, on closer inspection, the applicants were not, first of all, granted refugee status under the 1951 Convention, but rather that of "protected person" under New Zealand domestic law. Secondly, the judges did not consider it necessary to evaluate the elements relating to the damages caused by climate change, since the familial conditions of the applicants were in themselves sufficient to integrate a situation of "humanitarian exceptionality" by virtue of which the *non refoulement* principle had to be activated. The plaintiffs, husband, and wife with two dependent children (as well as with important family ties consolidated for generations in New Zealand) obtained not to be transferred to Tuvalu under the New Zealand Immigration Act of 2009, a law that establishes, among other things, the prohibition of deportation for humanitarian reasons when the conditions of the asylum seeker are exceptional to the point of making the transfer to the country of origin unfair or unduly harsh.[30] In the case at hand, the Tribunal, having seen and jointly considered the elements presented by the applicants (i.e., strong integration in the host community, their role in maintaining the health of the husband's mother—a full resident

[28] See New Zealand Immigration and Protection Tribunal, Decision No. [2014] NZIPT 501370-371. https://www.refworld.org/pdfid/585152d14.pdf. Accessed 13 July 2018.

[29] The Immigration and Protection Tribunal (NZIPT) replaced the Refugee Status Appeals Authority (RSAA) in actuation of 2009 New Zealand's Immigration Act.

[30] 2009 New Zealand Immigration Act, section 207: "*The Tribunal must allow an appeal against liability for deportation on humanitarian grounds only where it is satisfied that, a) there are exceptional circumstances of a humanitarian nature that would make it unjust or unduly harsh for the appellant to be deported from New Zealand; and b) it would not in all the circumstances be contrary to the public interest to allow the appellant to remain in New Zealand*".

in New Zealand—and the serious economic and environmental difficulties in Tuvalu), concluded for the humanitarian relevance of the case.[31] Thus, in motivating their decision, the judges stressed that the transfer of the applicants would separate them from a part of their family that had resided regularly in New Zealand for generations, therefore resulting in an "unusually significant disruption" in their family relationships, moreover putting at risk the health of the applicant's mother.

With regard to climate change, while accepting in general terms that natural disasters can theoretically take on humanitarian significance, the Tribunal emphasized that the evidence produced in every specific individual case must demonstrate how and why climate change may be able to make the transfer of the *particular* applicant unfair or unduly harsh. Once this was declared, the judges promptly underlined that in the case at hand it was not necessary to evaluate these elements, since the applicants' family relationships constituted an element already sufficient to raise a question of "humanitarian exceptionality" under the New Zealand Immigration Act.[32] Therefore, even though taken into consideration, climate change was not at all the cause that made the applicants' requests successful.

Moreover, regardless of any other consideration, this case has found its own legal bases in provisions of domestic law, which, although indirectly linked to the protection of human rights as governed by international law, have their *raison d'etre* exclusively in New Zealand.[33] In reality, there are few states that have autonomously adopted provisions, in the name of the need to protect human rights, which can become relevant at an international level for people who also suffer from environmental

[31] See NZIPT's decision *sub* note 339, para. 31.

[32] Ibid., paras. 31, 32 and 33.

[33] In the Italian context, please refer to a Bologna Court of Appeal decision of February 2016, which confirmed the first instance judgment that recognized a right to humanitarian protection to a Pakistani citizen suffering from frequent environmental disasters in his origin country. The Court of Appeal, after having underlined that there is to date no exhaustive and exhaustive list of the humanitarian reasons that allow the guarantee of a residence permit, and that these have therefore to be assessed case by case by competent courts (that must consider all relevant elements), established that "*nel caso di specie la particolare situazione dell'appellato, vittima di eventi alluvionali che lo hanno privato di ogni bene materiale e di ogni affetto nel paese di provenienza dove non è neppure possibile il ricorso alle autorità locali, sia meritevole della misura della protezione umanitaria*". Cf. Corte di Appello di Bologna, 19 February 2016 (N° 524/2016).

degradation associated with global warming. More in depth, a significant regulatory evolution is certainly that implemented in Sweden, where asylum law extends the protection regime also to people unable to return to their country due to an environmental disaster.[34] To date, this provision has not yet been applied and it remains to be seen, therefore, whether it is possible to apply it to gradual disasters and, above all, whether it can become relevant also with regard to cases where an internal migration within the affected state is in fact possible (the so-called reasonable internal flight alternative we talked of in Chapter 5). The Swedish law expressly provides that the protection can be extended to people *who cannot return* to their country of origin due to environmental disaster. Nevertheless, despite the obvious limitations, this provision can be considered as a starting point with a view to protecting environmental migrants, at least in the category of national legal orders.[35] Similar rules are also to be found in the Finnish legal system, where protection can be granted on humanitarian grounds if the person cannot return to his/her country due to an environmental catastrophe,[36] and in Argentina,[37] where the immigration law guarantees residence to individuals unable to return home due

[34] See Aliens Act Sweden (2005), chap. 4, sec. 2a: "In this Act, a 'person otherwise in need of protection' is an alien who in cases other than those referred to in Sections 1 and 2 is outside the country of the alien's nationality, because he or she [...] 2. is unable to return to the country of origin because of an environmental disaster".

[35] It is necessary to underline that internal law cannot in any case have any impact, if not at a comparative level, on Australian or New Zealand law and, therefore, on those legal systems that more than any other have to face the challenge of migrations originating from insular micro-states. In this terms, and with reference to Swedish Law and the OUA Convention for instance, has expressed itself the RRTA: "While it may be true that these developments have occurred elsewhere in consideration of human flight, the Tribunal is bound to apply the law as it currently stands in Australia". See Decision No. 0907346 [2009] RRTA 1168, 10 December 2009, para. 53. https://www.refworld.org/pdfid/4b8fdd952.pdf. Accessed 12 June 2017.

[36] Aliens Act Finland (301/2004).

[37] Decreto nr. 616/2010, *Boletìn Oficial de la Republica Argentina* nr. 31.898 (6 May 2010), Aricle 24.

to environmental or natural disasters.[38] However, even in these contexts it remains to be seen how the provisions can relate to climate change.[39]

In conclusion, therefore, complementary protection as currently structured and applied is unsuitable for protecting forced climate migrants as a category *per se*. Moreover, these systems exist in order to protect certain persons in the presence of particular individual situations that are unacceptable from a human rights protection point of view, and not to protect entire predetermined categories. In this sense, the difficult applicability of complementary protection regimes with respect to the phenomenon of climate-induced migration should not come as a surprise: the need for an imminent danger perceived by the applicants, as well as the exceptional and serious nature of their individual situation, eventually represent *criteria* by which complementary protection, where envisaged, could be applied only and exclusively to very few individuals because of completely peculiar individual circumstances. The New Zealand case mentioned above seems to confirm this approach.

In a nutshell, complementary protection frameworks are—through the obligation of protection from inhuman and degrading treatments—equipped with mechanisms and *criteria* that are potentially sufficient to

[38] A brief reference to the US Temporary Protection Status (TPS) mechanisms is useful. The TPS allows the Attorney General to activate protection measures for the citizens of a foreign country that is subject to a natural disaster. The TPS system grants facilitated access to labor market and impedes temporarily (for a period of 6 up to 16 months, that is however extendable) the repatriation of the citizens of the elected country that were already present on the territory of the United States at the time of the activation of the mechanism. What is interesting is that assistance is granted depending on the objective conditions of the origin country and not on the individual situation of the person. For further information, see Sighetti, Ester, & Wasem (2015). Temporary Protected Status: Current Immigration Policy and Issues, Congressional Research Service. https://www.everycrsreport.com/files/20150902_RS20844_9baedc8b7abaf46636f8ba4a179e745d9940a60a.pdf. Accessed 10 April 2019. See also Sciaccaluga (2017). Sudden-Onset Disasters, Human Displacement, and the Temporary Protection Directive: Space for a Promising Relationship? In Bruno, Palombino, & Rossi (Eds.), *Migration and the Environment: Some Reflections on Current Legal Issues and Possible Ways Forward*. Rome: CNR Edizioni, pp. 75 ff.

[39] See, for instance, Xing-Yin Ni (2005) (op. cit.), p. 357: "Whether refugee protection could actually be stretched far enough to apply to climate-induced migrants, however, has yet to be seen. A sudden-onset natural disaster may trigger protection under Swedish law or qualify as a circumstance 'seriously disturbing public order' under one of the regional treaties. An asylum seeker fleeing a slow-onset phenomenon, however, such as the sea-level rise in Kiribati, would be unlikely to meet even an expanded definition of refugee".

guarantee international protection only to some people forced to emigrate because of climate change, especially where their individual situation turns out to be exceptionally and extremely serious. It remains to be discovered, again, how the interested courts will rule on the matter. In the medium term, that is when global warming will have exacerbated certain environmental situations to the point of effectively rendering the permanence of some in their countries of origin inhuman and degrading, it seems likely that more and more appeals will begin to make their way in national and regional human rights protection systems, an eventuality that, in the long run, will throw light on the topic discussed here.

CHAPTER 12

Conclusion

This book sets itself the aim of conceptualizing the macro-category of "climate refugees" in a manner that is useful to international protection studies. Through an analysis of the interplay between climate change and human migration and of the various definitions and classifications proposed over the years to define "climate refugees", Part I shed light on what we called the "climate refugees cauldron". We dissected this confused and often incoherent set in which one may easily throw in any person whose migration—be it internal or international, voluntary or forced—is somehow linked to anthropogenic global warming. Relying on the functioning and rationale of international protection systems, this book argues that only a decidedly circumscribed subcategory of "climate refugees" should in fact be protected in a third country after a cross-border migration. To define this subcategory, we chose to use the term "forced climate migrants", who we identify as persons who are forced to leave the entirety of their national territory because this has become (or is about to become) uninhabitable due to climate change-driven environmental degradation. "Traditional" political refugees, as understood under the 1951 Refugee Convention milestone, are to be protected by a third state when their bond of trust with their country of origin is severed and when they consequently risk being persecuted (i.e., seeing their basic human rights seriously violated) if they stay in their home country. That

G. Sciaccaluga, *International Law and the Protection of "Climate Refugees"*,
https://doi.org/10.1007/978-3-030-52402-9_12

is why and when international law kicks in and grants them international protection through the *non refoulement* principle.

The basic idea underpinning this book follows the same line of thought with regard to climate-related migrations: only forced climate migrants should be protected in a third country. And this should occur when a person is not able to avail him/herself with the protection of the home country in relation to the enjoyment of his/her core, fundamental rights. Of course, the underlying causes that lead a political refugee and a forced climate migrant to leave their countries are different: the first is pushed by the persecution enacted (or permitted) by his or her state, the second because his/her state cannot cope with climate-driven environmental degradation that makes it impossible to live in tolerable life conditions. Attributing to climate change the inability of a migrant to enjoy basic human rights in his own national territory is, on the one hand, a limitation from an international protection viewpoint, since classic international protection systems do not hold environmental factors as capable of activating the *non refoulement* principle; on the other, it opens however an interesting avenue for the evolution of international law.

In Part II, the book delved into the analysis of existing international obligations with regard to climate change and human rights law, to bridge these two branches and to provide a systemic rationale that explains why states and the international community are bound to predispose protection measures for forced climate migrants. First, we looked at international climate change law and observed that its gradual evolution over the years is reducing more and more the plausibility of arguing that climate change mitigation and adaptation measures are not imposed by the law, be it in application of the UNFCCC and its implementation agreements or in application of customary international environmental law. The ever-growing and undisputed scientific certainty on the human origin of climate change, in association with the rising awareness on its disastrous transnational effects, has led to the adoption of the Paris Agreement, whose success or failure in the following decades will essentially depend on the assumption that "we are all in this together", and that each and every international law subject must do its best to avert the direst consequences of global warming. We therefore argued that international climate change law has evolved up to the point of imposing at minimum proportional national greenhouse gas emission cuts and climate change adaptation policies to every state. We also explained how these obligations reflect the rise of a new international custom, which requests states to act in response to climate change beyond treaty law. Furthermore, the final goal of safeguarding our planet's climate stability requires that similar obligations stand at an *erga omnes* level. In fact, if the Paris Agreement's 2°C goal will not be met, there is a serious risk of seeing the international order as we know it today drastically transformed. The climate stability that has characterized the last 10,000 years, since the beginning of the Holocene, is today in danger, with the Anthropocene that is dangerously elbowing to make room. In this author's opinion, relying also on what was acknowledged by the International Court of Justice with regard to the importance of planetary environmental protection, our climate stability is a "good" that is essential to the international community as a whole, a "good" on which the functioning (if not even the existence) of the international order actually depends. If this is the case, (inter)national climate change mitigation and adaptation measures are *erga omnes* obligations, since they aim at safeguarding a value that is essential to the entire international (legal) system. This entails that every state is nowadays bound, at least, to progressively reducing its greenhouse gas emissions and to adapt to the inevitable consequences of climate change.

We then wondered whether under these obligations, and in connection with human rights protection ones, there is an international duty to predispose protection measures for forced climate migrants. The rights to life, health, and access to basic livelihoods are non-derogable, inalienable, and universal. When their enjoyment in a certain political/territorial landscape is in danger, and forces a person to leave his/her country, international protection—be it through refugee law or complementary protection frameworks—requests that the international community (or, better said, the receiving state) becomes responsible for that person. The very same rationale should apply when a person flees his or her origin country because it is not possible to enjoy basic human rights there anymore. The 2020 UN HCR's decision concerning the *Teitiota case* has stated in fact the same: where a person risks serious violations of basic human rights if repatriated to a country heavily plagued by climate change, *non refoulement* obligations should apply.

When forced climate migrants are at play, the human rights obligations that "weigh" on states' shoulders are moreover reinforced by climate change obligations. If a political refugee is to be protected by the international community and a third state when persecuted by his or her own state, the same protection duty should be even more binding when a similar level of suffering is attributable to the international community as a whole. This means that every state, independently from its contribution to climate change, should offer protection to cross-border migrants when they can prove that a return to their country of origin would expose them to death or inhuman or degrading treatment because of climate change-driven ecosystemic degradations.

Of course, such a duty is not clearly reflected yet in existing international law instruments and case law, which by their very nature are anchored in a post-Holocaust system of values, where climate change and planetary environmental degradation play(ed) a decidedly marginal role. A gradual evolution of our legal system toward the recognition of international protection rights to forced climate migrants is however emerging, and the rationale described in this book will hopefully make its small contribution.

In fact, the embryos or seeds of this evolution are already visible. Part III of the book addressed them to understand where and how international law is and will be evolving in this sense. The progress made under the UNFCCC's umbrella in the last ten years, the success of intergovernmental dialogues such as the Nansen Initiative, and the recent adoption of the Global Compacts on Refugees and on Migration show that the international community and international law have started to address climate change-related human mobility in a quite exhaustive manner. We saw that the UNFCCC is probably the best *forum* where to gradually predispose studies, guidelines, and in the future maybe even binding rules in relation to the protection of forced climate migrants. We also saw that the GCR, given its recognition of climate change as a driver of refugee movements and of the need to fill in existing legal gaps to fulfill human rights obligations, has the potential to stimulate further developments in this sense, be it at the international, regional, or simply national level. The GCM as well comes as a promising step. Its explicit reference to climate and environment-related migrations and its clear call for implementing both intergovernmental preventive migration policies and protection instruments for migrants compelled to leave their countries witnesses that the international community is moving toward the widespread belief that action is to be undertaken to protect forced climate migrants.

With a view also to voluntary and preventive climate-related migration, we saw both in Part I and Chapter 10 that the whole "climate refugees" phenomenon (if considered only in its cross-border dimension) in fact presents two interlinked challenges: on the one hand, there is the need to predispose protection mechanisms for those who are forced to migrate because of climate change, on the other, there is the need to manage the increasing numbers of voluntary migrants who are (and will) be leaving their countries also because of climate change. Whereas this book has amply focused on the "protection" challenge of the phenomenon, the "migration management" one is equally (if not more) important, because it deals with far larger numbers of individuals. Hence, there is the urgent need to create and implement dedicated regular and preemptive migration pathways where possible. The GCM openly calls states in this respect. In fact, without a long-sighted management of climate-induced migration, international tensions concerning this sensitive topic appear destined to degenerate. The need to further study how regular migration pathways can be implemented to avert future uncontrollable "tides" of climate-induced migrants is crucial. Clearly, this will take time. This is because the GCM is a universal, soft law tool, which by its very nature is essentially rhetoric and exhortative when it comes to effectively directing state policies. At the same time, the GCM is groundbreaking, because for the first time in history states have gathered to create an all-comprehensive set of guidelines concerning global migration challenges. It will therefore be important to clarify how this instrument can and will have a concrete effect on the international management of climate-related migration in the decades to come.

In the last chapter, we then turned to complementary protection systems to understand whether and to what extent these are equipped to evolve to cover forced climate migrants under their umbrella. There appears to be some room for interpretation in this regard. Forced climate migrants may potentially be included among those who must not to be repatriated in order to protect their fundamental human rights and their dignity. Future studies and case law, hopefully relying on what recently established by recently the UN Human Rights Committee, could shed more light on this possibility. We saw that *non refoulement* applies in contexts of individual socio-economic distress in the country of origin only in extreme and exceptional circumstances. This raises therefore significantly the threshold beyond which a forced climate migrant could be covered by complementary protection frameworks. However, this is precisely the whole rationale that underlies international protection frameworks: foreigners are to be protected under international law in a third country only and insofar they risk being deprived of their core human rights in their home country. Forced climate migrants should be no exception in this sense, and if we start considering them as the only subcategory within the "climate refugees cauldron" that should effectively be protected abroad, they are more likely to be granted in practice the international protection they are due.

This approach clearly circumscribes the category of those who should be internationally protected as a result of climate-related migrations, but it is precisely because of this narrowing down that it can favor the development of protection measures for this category. This book delved into the principle that explains why people forced to leave their countries due to climate change should be protected under international law. It remains to be researched in the years ahead, whether and how this principle will be applied in practice.

Bibliography

Afifi et al. (2012). Climate Change, Vulnerability and Human Mobility: Perspectives of Refugees from the East and Horn Africa. Report. UNHCR & UNU-EHS. https://www.unhcr.org/protection/environment/4fe8538d9/climate-change-vulnerability-human-mobility-perspectives-refugees-east.html. Accessed 14 June 2019.

Albert et al. (2016). Interactions Between Sea-Level Rise and Wave Exposure on Reef Island Dynamics in The Solomon Islands. *Environmental Research Letters*, 11, pp. 1 ff.

Amnesty International (2020). UN Landmark Case for People Displaced by Climate Change. https://www.amnesty.org/en/latest/news/2020/01/un-landmark-case-for-people-displaced-by-climate-change/. Accessed 16 February 2020.

Apap (2019). The Concept of 'Climate Refugee'. Towards a Possible Definition. Briefing. EPRS | European Parliamentary Research Service. https://www.europarl.europa.eu/RegData/etudes/BRIE/2018/621893/EPRS_BRI(2018)621893_EN.pdf. Accessed 18 December 2019.

Atapattu (2016). *Human Rights Approaches to Climate Change: Challenges and Opportunities*. New York: Routledge.

Atapattu (2018). A New Category of Refugees? 'Climate Refugees' and Gaping Hole in International Law. In Behrman & Kent (Eds.), *'Climate Refugees' Beyond the Legal Impasse?* New York: Routledge, pp. 34 ff.

Bagliani & Dansero (2011). *Politiche per l'ambiente. Dalla natura al territorio*. Novara: Utet.

G. Sciaccaluga, *International Law and the Protection of "Climate Refugees"*,
https://doi.org/10.1007/978-3-030-52402-9

Baker & McKenzie (2016). The Paris Agreement Putting the First Universal Climate Change Treaty In Context. https://www.lexology.com/library/detail.aspx?g=7f89d079-a97c-4c87-99ba-16f3c9ff5fdf. Accessed 13 January 2018.

Bakewell (2008). Keeping Them in Their Place: The Ambivalent Relationship Between Development and Migration in Africa. *Third World Quarterly*, 7 (29), pp. 1341 ff.

Barnett & Webber (2010). Migration as Adaptation. In McAdam (Ed.), *Climate Change and Displacement, Multidisciplinary Perspectives*. Oxford: Hart Publishing.

Bates (2002). Environmental Refugees? Classifying Human Migrations Caused by Environmental Change. *Population and Environment*, 5, pp. 466 ff.

Bawden (2015, March 2). Climate Change Key in Syrian Conflict—And It Will Trigger More War in Future. *The Independent*. https://www.independent.co.uk/news/world/middle-east/climate-change-key-in-syrian-conflict-and-it-will-trigger-more-war-in-future-10081163.html. Accessed 10 June 2019.

Beck (2011). *Disuguaglianza senza confini*. Bari: Editori Laterza.

Behrman & Kent (Eds.) (2018). *'Climate refugees' Beyond the Legal Impasse?* New York: Routledge.

Bergoglio (2015). *Lettera enciclica – Laudato si' – del Santo Padre Francesco sulla cura della casa comune*. Vatican City: Libreria Editrice Vaticana.

Bettini (2013). Climate Barbarians at the Gate? A Critique of Apocalyptic Narratives on 'Climate Refugees'. *Geoforum*, 45, pp. 63 ff. https://www.sciencedirect.com/science/article/abs/pii/S0016718512001947?via%3Dihub. Accessed 12 January 2020.

Bilsborrow (1992). Rural Poverty, Migration and the Environment in Developing Countries: Three Case Studies, World Bank Policy Research Working Paper 1017. World Bank. http://documents.worldbank.org/curated/en/77769146876738 6516/pdf/multi0page.pdf. Accessed 14 May 2019.

Black (1998). *Refugees, Environment and Development*. New York: Harlow Longman.

Bodansky (2015). Reflections on the Paris Conference. *Opinio Juris*. http://opiniojuris.org/2015/12/15/reflections-on-the-paris-conference/. Accessed 14 January 2019.

Bohle, Cannon, Hugo & Ibrahim (Eds.) (1991). *Famine and Food Security in Africa and Asia: Indigenous Response and External Intervention to Avoid Hunger*. Bayreuth: Bayreuther Geowissenschaftilche Arbeiten.

Boyd (2012). *The Environmental Rights Revolution*. Vancouver: UBC Press.

Bratspies & Miller (2006). *Transboundary Harm in International Law: Lessons from the Trail Smelter Arbitration*. Cambridge: Cambridge University Press.

Brown & Oli (2008). Migration and Climate Change. International Organization for Migration: Research Series No. 31 https://publications.iom.int/system/files/pdf/migration_and_environment.pdf. Accessed 13 May 2019.

Brown (2007). Climate Change and Forced Migration: Observations, Projections and Implications. Human Development Office Occasional Paper. UNEP, p. 7. https://www.iisd.org/pdf/2008/climate_forced_migration.pdf. Accessed 15 April 2019.

Brown (2016). Europe's Refugee Crisis: A Taste of Things to Come? *UNEP's Disaster and Conflicts Subprogramme.* https://medium.com/@UNEP/europe-s-refugee-crisis-a-taste-of-things-to-come-72a5d4079cb4. Accessed 30 May 2019.

Byravan & Rajan (2006). Providing New Homes for Climate Change Exiles *Climate Policy*, 6, pp. 247 ff.

Capra (1975). *The Tao of Physics.* Boston: Shambhala.

Castles (2010). Afterword: What Now? Climate-Induced Displacement After Copenhagen. In McAdam (Ed.), *Climate Change and Displacement, Multi-disciplinary Perspectives.* Oxford: Hart Publishing.

Celle (2011). The Right of Asylum: Towards a Common European System. Impact on the Mediterranean Area, in Bottaro Palumbo & Danisi (Eds.), *Civil Rights Protection and the Rights of Migrants in the Framework of the Mediterranean Cooperation.* Genoa: Genova University Press.

Christian Aid (2007). Human Tide: The Real Migration Crisis. https://www.christianaid.org.uk/sites/default/files/2017-08/human-tide-the-real-migration-crisis-may-2007.pdf. Accessed 13 April 2019.

Cooper (1998). Environmental Refugees: Meeting the Requirements of the Refugee Definition. *New York University Environmental Law Journal*, 6.

Craven (2007). The Violence of Dispossession: Extraterritoriality and Economic, Social and Cultural Rights, in Baderin & McCorquodale (Eds.), *Economic Social and Cultural Rights in Action.* Oxford: Oxford University Press, Oxford, pp. 71 ff.

Curtain et al. (2016). Labour Mobility: The Ten Billion Dollar Prize. Report. World Bank. http://documents.worldbank.org/curated/en/171661503669342316/pdf/119105-PUB-PUBLIC-ADD-SERIES-pplabourmobilitybackgroundfinal.pdf. Accessed 14 February 2020.

Daily (Ed.) (1997). *Nature's Services: Societal Dependence on Natural Systems.* Washington, DC: Island Press.

D'Angelo (2016, December 2). A Louisiana Tribe is Now Officially a Community of Climate Refugees. *Huffington Post Hawaii.* https://www.huffpost.com/entry/climate-refugees-louisiana_n_56bbd5efe4b0c3c550501784?guccounter=1&guce_referrer=aHR0cHM6Ly93d3cuZ29vZ2xlLmNvbS8&guce_referrer_sig=AQAAAAcHPSHZKmfQvaam-92okhuewL22cmI0xVjAooX_VgJLoxMLeMitEZ8vx9OpvK-dqLh3DvXH6VQ6ojgsvgXl0MhrbkMWCxjTMXDXzyFM18gDaDQnpWk43ZDTMkPjgCK61_7SELfa-dTMlVe5962oletUa7IvGr9kf_XI85s8xBiu. Accessed 14 April 2019.

Déleage (1994). *Storia dell'ecologia. Una scienza dell'uomo e della natura.* Naples: Cuen.

Diamond (2005). *Collasso. Come le società scelgono di morire o vivere.* New York: Viking Press.

DiChristopher (2018, July 19). Judge Throws Out New York City's Climate Change Lawsuit Against Five Major Oil Companies. *CNBC.* https://www.cnbc.com/2018/07/19/judge-tosses-nycs-climate-change-lawsuit-against-5-big-oil-companies.html. Accessed 20 July 2019.

Doelle (2015). The Paris Agreement: Historic Breakthrough or High Stakes Experiment? https://papers.ssrn.com/sol3/papers.cfm?abstract_id=2708148. Accessed 14 January 2018.

Doherrty & Slezak (2017, August 10). Australia Faces Potentially Disastrous Consequences of Climate Change, Inquiry Told. *The Guardian.* https://www.theguardian.com/environment/2017/aug/11/australia-potentially-disastrous-consequences-of-climate-change-inquiry-told. Accessed 12 January 2019.

Donald Hughes (1975). *Ecology in Ancient Civilizations.* Albuquerque: University of New Mexico Press.

EACH-FOR (2009). Environmental Change and Forced Migration Scenarios, Synthesis Report. http://rosamartinez.org/wp-content/uploads/2015/11/Migraciones-y-Cambio-Climatico_EACHFOR.pdf. Accessed 2 September 2018.

El-Hinnawi (1985). Environmental Refugees. Report. UNEP. https://digitallibrary.un.org/record/121267. Accessed 12 February 2020.

European Commission (2016). The Road from Paris: Assessing the Implications of the Paris Agreement and Accompanying the Proposal for a Council Decision on the Signing, on Behalf of the European Union, of the Paris Agreement Adopted Under the United Nations Framework Convention on Climate Change. Communication. (COM(2016) 110 final).

Falstron (2001). Stemming the Flow of Environmental Displacement: Creating a Convention to Protect Persons and Preserve the Environment. *Colorado Journal of Environmental Law and Policy*, 15, pp. 1 ff.

Farand (2017, April 1). Nine-Year-Old Girl Files Lawsuit Against Indian Government Over Failure to Take Ambitious Climate Action. *The Independent.* https://www.independent.co.uk/environment/nine-ridhima-pandey-court-case-indian-government-climate-change-uttarakhand-a7661971.html. Accessed 23 January 2020.

Foster (2007). *International Refugee Law and Socio-Economic Rights: Refugee from Deprivation.* Cambridge: Cambridge University Press.

Fountain (2015, March 2). Researchers Link Syrian Conflict to a Drought Made Worse by Climate Change. *The New York Times.* https://www.nytimes.com/2015/03/03/science/earth/study-links-syria-conflict-to-drought-caused-by-climate-change.html. Accessed 9 June 2019.

Franca (2016). Climate Change and Interdependent Human Rights to Food, Water and Health: The Contest Between Harmony and Invention. In Quirico & Boumghar (Eds.), *Climate Change and Human Rights: An International and Comparative Law Perspective*. New York: Routledge, pp. 89–103.

Francioni (2014). Comments on the Preamble and on Principle 1. In Viñuales (Ed.), *The Rio Declaration on Environment and Development: A Commentary*. Oxford: Oxford University Press.

Francioni & Quirico (2016). Untying the Gordian Knot. In Quirico & Boumghar (Ed.), *Climate Change and Human Rights*. New York: Routledge, pp. 147 ff.

Friends of Earth in Australia (2007). A Citizen's Guide to Climate Refugees. https://www.safecom.org.au/pdfs/foe-climate-citizens-guide.pdf. Accessed 12 May 2019.

Fumagalli (2016). Illecito e Responsabilità. In Carbone et al. (Eds.), *Istituzioni di Diritto Internazionale*. Turin: Giapichelli, pp. 269 ff. See also: Crawford et al. (2010). *The Law of International Responsibility*. Oxford: Oxford University Press.

Gemenne (2011). Climate-Induced Displacement in a 4°C+ Degree World. *Philosophical Transactions of the Royal Society A*, 182.

Gemenne (2017). The Refugees of the Anthropocene. In Mayer & Crépeau (Eds.), *Research Handbook on Climate Change, Migration and the Law*. Cheltenham: Elgar, pp. 394 ff.

Gervasi (2016). Rilievi critici sull'Accordo di Parigi: le sue potenzialità e il suo ruolo nell'evoluzione dell'azione internazionale di contrasto al cambiamento climatico. *La Comunità Internazionale*, pp. 21 ff.

Giovannini (2017, November 6). Un miliardo di rifugiati climatici entro il 2050. *La Stampa*. https://www.lastampa.it/tuttogreen/2017/11/06/news/un-miliardo-di-rifugiati-climatici-entro-il-2050-1.34380312. Accessed 13 January 2020.

Giuliano, Scovazzi & Treves (1983). *Diritto internazionale. Vol. 2: Gli aspetti giuridici della coesistenza degli Stati*. Milan: Giuffré.

Goudarzi (2016). As Earth Warms, the Diseases That May Lie Within Permafrost Become a Bigger Worry *Scientific American*. https://www.scientificamerican.com/article/as-earth-warms-the-diseases-that-may-lie-within-permafrost-become-a-bigger-worry/. Accessed 12 May 2019.

Henkin (1990). *The Age of Rights*. New York: Columbia University Press.

Hens (2012). Environmentally Displaced People. *Encyclopedia of Life Support Systems, Area Studies Africa (Regional Sustainable Development Review)—Vol. II*. http://web.mnstate.edu/robertsb/308/environmentally%20displaced%20people.pdf. Accessed 14 May 2019.

Hillary (Ed.) (2000). *Ecology 2000: The Changing Face of Earth, Joseph*. New York: Beaufort Books.

Homer-Dixon & Percival (1996). *Environmental Scarcity and Violent Conflict: Briefing Book*. Toronto: University of Toronto and the American Association for the Advancement of Science.

Hoss (2005). State Responsibility, Liability and Environmental Protection. In Wolfrum, Langenfeld & Minnerop (Eds.), *Environmental Liability in International Law: Towards a Coherent Conception*. Berlin: Erich Schmidt Verlag, pp. 455–494.

Huang (2015). The 2015 Climate Agreement: Key Lessons Learned and Legal Issues on the Road to Paris. https://papers.ssrn.com/sol3/papers.cfm?abstract_id=2724109. Accessed 14 April 2019.

Huber & Gulledge (2011). Extreme Weather and Climate Change: Understanding the Ling and Managing the Risk. Center for Climate and Energy Solutions. https://www.c2es.org/site/assets/uploads/2011/12/white-paper-extreme-weather-climate-change-understanding-link-managing-risk.pdf. Accessed 13 October 2019.

Hugo (1995). Environmental Concerns and International Migration. *International Migration Review*, pp. 105 ff.

Hugo (1996). Environmental Concerns and International Migration. *International Migration Review*, 30, Ch. 6.

Hugo (2010). Climate Change-Induced Mobility and Existing Migration Regime in Asia and the Pacific. In McAdam (Ed.), *Climate Change and Displacement, Multidisciplinary Perspectives*. Oxford: Hart Publishing.

IASC Task Force on Climate Change (2010). Quick guide to Climate Change Adaptation. IASC. https://interagencystandingcommittee.org/system/files/legacy_files/IASC%20Quick%20Guide%20Adaptation.pdf. Accessed 12 May 2019.

IDMC (2015). People Displaced by Disasters. Global Estimates 2015. Report. IDMC. https://www.internal-displacement.org/sites/default/files/inline-files/20150713-global-estimates-2015-en-v1.pdf. Accessed 18 March 2020.

Incropera (2016). *Climate Change: A Wicked Problem*. Cambridge: Cambridge University Press.

IOM (2019). Environmentally-Induced Population Displacements and Environmental Impacts Resulting from Mass Migration (p. 4). International Symposium, IOM. https://publications.iom.int/system/files/pdf/environmentally_induced.pdf. Accessed 12 December 2019.

IPCC (1990). First Assessment Report. Climate Change: The IPCC Impacts Assessment (1990), Policymakers' Summary. https://www.ipcc.ch/site/assets/uploads/2018/05/ipcc_90_92_assessments_far_full_report.pdf. Accessed 15 March 2020, p. 3.

IPCC (2007). Climate Change 2007: Synthesis Report, pp. 30 and 53. https://www.ipcc.ch/site/assets/uploads/2018/02/ar4_syr_full_report.pdf. Accessed 12 May 2019.

IPCC (2014). Summary for Policymakers. In *Climate Change 2014: Impacts, Adaptation, and Vulnerability. Part A: Global and Sectoral Aspects*. Cambridge: Cambridge University Press, p. 20. https://www.ipcc.ch/site/assets/uploads/2018/02/WGIIAR5-PartA_FINAL.pdf. Accessed 15 August 2019.

IPCC (2014). Summary for Policymakers. In *Climate Change 2014: Impacts, Adaptation, and Vulnerability. Part A: Global and Sectoral Aspects*. Work of Working Group II of the Fifth Intergovernmental Panel on Climate Change Report. Cambridge: Cambridge University Press, p. 20.

IPCC (2016). Fifth Report, Working Group I, The Scientific Base. IPCC. https://archive.ipcc.ch/ipccreports/tar/wg1/016.htm. Accessed 13 January 2020.

Jacobson (1988). Environmental Refugees. A Yardstick of Habitability. *Bulletin of Science, Technology and Society* 8, pp. 257 ff.

Jastram (2014). Warm World, Cold Reception: Climate Change, National Security and Forced Migration. *Veromont. Journal of Envtl. Law*, 2014, pp. 752 ff.

Kaiser (2011). *How the Hippies saved Physics*. London: W. W. Norton.

Kälin & Schrepfer (2012). Protecting People Crossing Borders in the Context of Climate Change: Normative Gaps and Possible Approaches. UNHCR Legal and Protection Research Series, PPLA/2012/01 https://www.unhcr.org/4f33f1729.pdf. Accessed 12 December 2016.

Kaminski (2019, December 9). Carbon Majors Can Be Held Liable for Human Rights Violations, Philippines Commission Rules. *Climate Liability News*. https://www.climateliabilitynews.org/2019/12/09/philippines-human-rights-climate-change-2/. Accessed 20 March 2020.

Keane (2004). The Environmental Causes and Consequences of Migration: A Search for the Meaning of 'Environmental Refugees'. *Geo. Int'l Law Review*, 2, pp. 210 ff.

Kelley et al. (2015). Climate change in the Fertile Crescent and Implications of the Recent Syrian Drought. *Proceedings of the National Academy of Science*, 11, pp. 3241 ff.

Kibreab (1997). Environmental Causes and Impact of Refugee Movements: A Critique of the Current Debate. *Disasters*, 21, pp. 20 ff.

Kiss & Shelton (2007). Strict Liability in International Environmental Law. In Nndiaye & Wolfrum (Eds.), *Law of the Sea, Environmental Law and Settlement of Disputes. Liber Amicorum Judge Thomas A. Mensah*. Leiden: Martinus Nijhoff Publishers, pp. 1131 ff.

Knox (2015). UN Human Rights Council's Report of the Independent Expert on the Issue of Human Rights Obligations Relating to the Enjoyment of a Safe, Clean, Healthy and Sustainable Environment. Compilation of Good Practices (A/HRC/28/61). Koslopova (2013). *Interstate Liability for Climate Change-Related Damage*. The Hague: Eleven.

Laczko & Agharzam (Eds.) (2009). *Migration, Environment and Climate Change: Assessing the Evidence*. Geneva: IOM.

Lazarus (2009). Super Wicked Problems and Climate Change: Restraining the Present to Liberate the Future. *Cornell Law Review*, 94, pp. 1153 ff.

Lenzerini (2009). *Asilo e diritti umani- Evoluzione del diritto d'asilo nel diritto internazionale*. Milan: Giuffré.

Lovelock (1979). *Gaia: A New Look at Life on Earth*. Oxford: Oxford University Press.

Martin (2017). Towards an Extension of Complementary Protection? In Mayer & Crépeau (Eds.), *Research Handbook on Climate Change, Migration and the Law*. Cheltenham: Elgar, pp. 449 ff.

Mayer (2012). 'Environmental Refugees'? A Critical Perspective on the Normative Discourse. Center for Sustainable Development Law. https://papers.ssrn.com/sol3/papers.cfm?abstract_id=2111825. Accessed 14 January 2016.

Mayer (2017). Climate Change, Migration and the Law of State Responsibility. In Mayer & Crépeau (Eds.), *Research Handbook on Climate Change, Migration and the Law*. Cheltenham: Elgar.

McAdam (2007). *Complementary Protection in International Refugee Law*. Oxford: Oxford University Press. Goodwin-Gill (1986). Non-refoulement and the New Asylum Seekers. *Virginia Journal of International Law*, 1986, pp. 897 ff.

McAdam (2011). Climate Change Displacement and International Law: Complementary Protection Standards. UNHCR Legal and Protection Policy Research Series, doc. PPLA/2011/03. https://www.unhcr.org/4dff16e99.pdf. Accessed 10 February 2018.

McAdam (2012). *Climate Change, Forced Migration, and International Law*. Oxford: Oxford University Press.

McAdam (Ed.) (2010). *Climate Change and Displacement, Multidisciplinary Perspectives*, Oxford: Hart Publishing.

McInerney-Lankford (2013). Human Rights and Climate Change. In Gerrard & Wannier (Eds.), *Threatened Island Nations: Legal Implications of Rising Seas and a Changing Climate*. New York: Cambridge University Press, pp. 229 ff.

McKie (2010, 28 November). Climate Change Will Cost a Billion People Their Homes, Says Report. *The Observer*. https://www.theguardian.com/environment/2010/nov/28/cancun-climate-summit-weather. Accessed 13 May 2019.

McNeill (2010). *Du Nouveau sous le Soleil: histoire de l'environnement mondial au XXe siècle*. Seyssel: Champ Vallon.

Merica (2015, November 4). Hillary Clinton: Climate Change Has Contributed to Refugee Crises, Including Syria. *CNN*. https://edition.cnn.com/2015/11/04/politics/hillary-clinton-climate-change-syria-refugee-crises/index.html. Accessed 13 March 2019.

Meron (1995). Extraterritoriality of Human Rights Treaties, *American Journal of International Law*, 89, pp. 78–82.

Merryman (1977). Comparative Law and Social Change: on the Origins, Style, Decline & Revival of the Law and Development Movement. *American Journal of Comparative Law*, 25, pp. 457–491.

Meze-Hausken (2000). Migration, Development and Environment: How Vulnerable Are People in Dryland Areas? A Case Study in Northern Ethiopia. *Mitigation and Adaptation Strategies for Global Change*, 5, 2000, pp. 379 ff.

Millennium Ecosystem Assessment (20015). *Ecosystems and Human Well-being: Synthesis*. Washington, DC: Island Press.

Morel (2014). *The Right Not to Be Displaced in International Law*. Cambridge: Intersentia.

Motzfeldt-Kravik (2018). The Security Council and Climate Change—Too Hot to Handle? *EJIL: Talk*. https://www.ejiltalk.org/the-security-council-and-climate-change-too-hot-to-handle/. Accessed 14 September 2019.

Munari (2016). The Perfect Storm on EU Asylum Law: The Need to Rethink the Dublin Regime. *Diritti Umani e Diritto Internazionale*, pp. 517 ff.

Munari & Schiano di Pepe (2012). *Tutela Transnazionale dell'Ambiente*. Bologna: il Mulino.

Myers (1993). Environmental Refugees in a Globally Warmed World. *Bioscience*, 43, pp. 752 ff.

Myers (2001). Environmental Refugees: A Growing Phenomenon of the 21st Century. *Philosophical Transactions of the Royal Society: Biological Sciences*, 357, pp. 609 ff.

Myers (2005). Environmental Refugees: An Emergent Security Issue, 13. Economic Forum, Prague, 23–27 May 2005. https://www.osce.org/eea/14851?download=true. Accessed 12 May 2018.

Myers & Kent (1995). Environmental Exodus: An Emergent Crisis in the Global Arena, Climate Institute, Washington, DC. https://portals.iucn.org/library/node/22380 accessed 17 March 2020.

Nansen Conference on Climate Change and Displacement in the 21st Century (2011). Chairperson's Summary. https://www.unhcr.org/4ea969729.pdf. Accessed 12 January 2020.

Nespor (2009). *Il Governo dell'Ambiente*. Milan: Garzanti.

Ntekangi (2014). *Vers un droit international des réfugiés écologiques*. Kinshasa: L'Harmattan.

Nuwer (2017, April 1). How Western Civilisation Could Collapse. *BBC Online*. https://www.bbc.com/future/article/20170418-how-western-civilisation-could-collapse. Accessed 13 June 2029.

Obergassel et al. (2016). Phoenix from Ashes—An Analysis of the Paris Agreement to the United Nations Framework, Wuppertal Institute for Climate,

Environment and Energy. https://epub.wupperinst.org/frontdoor/deliver/index/docId/6374/file/6374_Obergassel.pdf. Accessed 13 January 2018.

OCHA, IDMC & Norwegian Refugee Council (2009). Monitoring Disaster Displacement in the Context of Climate Change: Findings of a Study by the United Nations Office for the Coordination of Humanitarian Affairs and the International Displacement Monitoring Center. https://www.internal-displacement.org/sites/default/files/publications/documents/200909-monitoring-disaster-displacement-thematic-en.pdf. Accessed 14 November 2019.

Pall et al. (2011). Anthropogenic Greenhouse Gas Contribution to Flood Risk in England and Wales in Autumn 2000. *Nature*, 470 (7334), pp. 382 ff.

Parson (2017). Opinion: Climate Policymakers and Assessments Must Get Serious About Climate Engineering. *Proceedings of the National Academy of Sciences*, 35. https://www.pnas.org/content/pnas/114/35/9227.full.pdf. Accessed 4 August.

Parson (2017). Starting the Dialogue on Climate Engineering Governance: A World Commission. Fixing Climate Governance Policy Brief No. 8. Centre for International Governance Innovation. https://www.cigionline.org/publications/starting-dialogue-climate-engineering-governance-world-commission. Accessed 4 August 2019.

Pascoe (2015). Sailing the Waves on Our Own: Climate Migration, Self-determination and the Carteret Islands. *QUT Law Review*, 2, 2015, pp. 72 ff.

Pavoni (2014). *Interesse pubblico e diritti individuali nella giurisprudenza della Corte europea dei diritti umani*. Naples: Editoriale Scientifica.

Picone (2013). *Comunità Internazionale ed Obblighi Erga Omnes*. Naples: Jovene Editore.

Piguet (2008). Climate Change and Forced Migration, New Issues in Refugee Research, Research Paper n. 153. UNHCR, p. 8. https://www.unhcr.org/en-lk/47a316182.pdf. Accessed 13 May 2018.

Poon (2018). Drawing Upon International Refugee Law. The Precautionary Approach to Protecting Climate Change-Displaced Persons. In Behrman & Kent (Eds.), *'Climate Refugees' Beyond the Legal Impasse?* New York: Routledge, pp. 157 ff.

Prieur et al. (2013). Draft Convention on the International Status of Environmentally-Displaced Persons (Third Version). University of Limoges. https://cidce.org/wp-content/uploads/2016/08/Draft-Convention-on-the-International-Status-on-environmentally-displaced-persons-third-version.pdf. Accessed 12 January 2020.

Randall (2016). Syria and Climate Change: Did the Media Get It Right? *Climate and Migration Coalition*. http://climatemigration.org.uk/syria-and-climate-change-did-the-media-get-it-right-2/. Accessed 19 May 2019.

Renaud, Bogardi, Dun & Warner (2007). Control, Adapt or Flee: How to Face Environmental Migration? *InterSecTions Publication Series*, 5. UNU-EHS. https://collections.unu.edu/view/UNU:1859#:~:text=Control%2C%20adapt%20or%20flee%3A%20how%20to%20face%20environmental%20migration%3F,-Series%20Title&text=The%20complex%20subject%20of%20this,sudden%20onset%20of%20hazard%20events. Accessed 10 June 2019.

Report of the Task Force on Displacement, Advanced Unedited Version of 17 September 2018. https://unfccc.int/sites/default/files/resource/2018_TFD_report_17_Sep.pdf. Accessed 14 March 2020.

Romanin Jacur (2016). The Paris Agreement on Climate Change: Progresses Achieved and Challenges Ahead. *NCTM Newsletter*. https://www.nctm.it/news/articoli/the-paris-agreement-on-climate-change-progresses-achieved-and-challenges-ahead. Accessed 12 January 2018.

Schleussner et al. (2016). Armed-Conflict Risks Enhanced by Climate-Related Disasters in Ethnically Fractionalized Countries. *PNAS*, 33. https://www.pnas.org/content/pnas/113/33/9216.full.pdf. Accessed 14 October 2019.

Sciaccaluga (2015). La Corte Distrettuale dell'Aja e la lotta ai cambiamenti climatici: valori e limiti della politica climatica dell'Unione europea. *Eurojus.it Rivista*. http://rivista.eurojus.it/la-corte-distrettuale-dellaja-e-la-lotta-ai-cambiamenti-climatici-valori-e-limiti-della-politica-climatica-dellunione-europea/. Accessed 13 April 2019.

Sciaccaluga (2017). Sudden-Onset Disasters, Human Displacement, and the Temporary Protection Directive: Space for a Promising Relationship? In Bruno, Palombino & Rossi (Eds.), *Migration and the Environment: Some Reflections on Current Legal Issues and Possible Ways Forward*. Rome: CNR Edizioni, pp. 75 ff.

Secretariat of the Convention on Biological Diversity (2016). Managing Ecosystems in the Context of Climate Change Mitigation: A Review of Current Knowledge and Recommendations to Support Ecosystem-Based Mitigation Actions That Look Beyond Terrestrial Forests. CBD Technical Series No. 86. https://www.cbd.int/doc/publications/cbd-ts-86-en.pdf. Accessed 12 May 2019.

Sen (1993). Overcoming Global Hunger, Actions to Reduce Hunger World-Wide. Report. http://www.nzdl.org/gsdlmod?e=d-00000-00---off-0hdl--00-0----0-10-0---0---0direct-10---4-------0-1l--11-en-50---20-about---00-0-1-00-0--4----0-0-11-10-0utfZz-8-00&a=d&cl=CL2.22.2&d=HASH0144a39d21fe11a28a32cda9.12.fc. Accessed 10 March 2020.

Shacknove (1985). Who Is a Refugee? *Ethics*, 95, pp. 275 ff.

Shurke & Visentin (1992). The Environmental Refugee: A New Approach, Ecodecision. *Canada's Periodical on Refugees*, 1, pp. 1 ff.

Sighetti, Ester & Wasem (2015). Temporary Protected Status: Current Immigration Policy and Issues, Congressional Research Service. https://www.everycrsreport.com/files/20150902_RS20844_9baedc8b7abaf46636f8ba4a179e745d9940a60a.pdf. Accessed 10 April 2019.

Sinclair-Blakemore (2020). Teitiota v New Zealand: A Step Forward in the Protection of Climate Refugees Under International Human Rights Law? *Oxford Human Rights Hub Blog*. https://ohrh.law.ox.ac.uk/teitiota-v-new-zealand-a-step-forward-in-the-protection-of-climate-refugees-under-international-human-rights-law/. Accessed 13 March 2020.

Skogly (2006). *Beyond National Borders: States Human Rights Obligations in Development Cooperation*. Antwerpen and Oxford: Intersentia.

Spijkerboer (2002). Subsidiarity in Asylum Law: The Personal Scope of International Protection. In Bouteillet-Pacquet (Ed.), *Subsidiary Protection of Refugees in the European Union: Complementing the Geneva Convention?* Brussels: Bruylant, pp. 19 ff.

Stern (2007). *The Economics of Climate Change: The Stern Review*. Cambridge: Cambridge University Press.

Supran & Oreskes (2017). Assessing ExxonMobil's Climate Change Communications (1977–2014). *Environmental Research Letters*, 12 https://iopscience.iop.org/article/10.1088/1748-9326/aa815f/pdf. Accessed 13 March 2019.

Terminski (2012). Towards Recognition and Protection of Forced Environmental Migrants in the Public International Law. Refugee or IDPs Umbrella? https://papers.ssrn.com/sol3/papers.cfm?abstract_id=2029796. Accessed 14 September 2019.

The Lancet Countdown on Health and Climate Change (2017): From 25 Years of Inaction to a Global Transformation for Public Health. https://www.thelancet.com/journals/lancet/article/PIIS0140-6736(17)32464-9/fulltext. Accessed 13 May 2019.

The Nansen Initiative (2015). Agenda for the Protection of Cross-Border Displaced Persons in Context of Disasters and Climate Change Volume I, par. 14. https://nanseninitiative.org/wp-content/uploads/2015/02/PROTECTION-AGENDA-VOLUME 1.pdf. Accessed 19 March 2020.

Timberlake (1983). *Environmental Wars and Environmental Refugees: The Political Background to the Cartagena Convention*. London: Earthscan.

Todorov (2009). *La paura dei barbari: oltre lo scontro delle civiltà*. Milan: Garzanti.

Tomuschat (2010). The Right to Life: Legal and Political Foundations. In Tomuschat, Lagrange & Oeter (Eds.), *The Right to Life*. Leiden: Brill, pp. 3 ff.

UN Commission on Human Rights (2015). Human Rights Resolution 2005/16: Human Rights and Extreme Poverty (E/CN.4/RES/2005/16).

https://www.refworld.org/docid/45377c340.html. Accessed 10 March 2020.

UN Committee on Economic, Social and Cultural Rights (CESCR) (1999). General Comment No. 12: The Right to Adequate Food (Art. 11 of the Covenant). https://www.refworld.org/docid/4538838c11.html. Accessed 30 September 2019.

UN Committee on Economic, Social and Cultural Rights (CESCR) (1990). General Comment No. 3: The Nature of States Parties' Obligations (Art. 2, Para. 1, of the Covenant). https://www.refworld.org/docid/4538838e10.html. Accessed 28 March 2019.

UN Committee on Economic, Social and Cultural Rights (CESCR) (1999). General Comment No. 12: The Right to Adequate Food (Art. 11 of the Covenant). https://www.refworld.org/docid/4538838c11.html. Accessed 23 September 2019.

UN Committee on Economic, Social and Cultural Rights (CESCR) (2000). CESCR General Comment No. 14: The Right to the Highest Attainable Standard of Health (Art. 12). https://www.refworld.org/docid/4538838d0.html. Accessed 14 June 2019.

UN Committee on Economic, Social and Cultural Rights (CESCR) (2003). General Comment No. 15: The Right to Water (Arts. 11 and 12 of the Covenant). https://www.refworld.org/docid/4538838d11.html. Accessed 13 September 2019.

UN Committee on the Elimination of Racial Discrimination (CERD) (2010). Consideration of Reports Submitted by States Parties Under Article 9 of the Convention: Concluding Observations of the Committee on the Elimination of Racial Discrimination: Japan. https://www.refworld.org/docid/4c15ef6b2.html. Accessed 10 May 2019.

UN Committee on the Rights of the Child (CRC) (2005). General Comment No. 6 (2005): Treatment of Unaccompanied and Separated Children Outside Their Country of Origin. https://www.refworld.org/docid/42dd174b4.html. Accessed 12 February 2020.

UN Human Rights Committee (HRC) (1986). CCPR General Comment No. 15: The Position of Aliens Under the Covenant. https://www.refworld.org/docid/45139acfc.html. Accessed 10 January 2018.

UN Human Rights Committee (HRC) (2004). General Comment No. 31: The Nature of the General Legal Obligation Imposed on States Parties to the Covenant. https://www.refworld.org/docid/478b26ae2.html. Accessed 20 March 2020.

UN Human Rights Committee (HRC) (2020). CCPR General Comment No. 36: Article 6 (Right to Life). https://www.refworld.org/docid/5e5e75e04.html. Accessed 19 May 2020.

UN Human Rights Council (2008). Human Rights and Climate Change. https://ap.ohchr.org/documents/E/HRC/resolutions/A_HRC_RES_7_23.pdf. Accessed 15 March 2019.

UN Human Rights Council (2008). Report of the Special Rapporteur on the Right to Food, Jean Ziegler. https://www.refworld.org/docid/47c3dbe82.html. Accessed 15 January 2020.

UN Human Rights Council (2009). Report of the Office of the United Nations High Commissioner for Human Rights on the Relationship Between Climate Change and Human Rights. A/HRC/10/61. https://www.refworld.org/docid/498811532.html. Accessed 7 September 2019.

UN Office for Disaster Risk Reduction (2009). UNISDR Terminology on Disaster Risk Reduction. https://www.unisdr.org/files/7817_UNISDRTerminologyEnglish.pdf. Accessed 15 May 2019.

UN OHCHR (2014). Mapping Human Rights Obligations Relating to the Enjoyment of a Safe, Clean, Healthy and Sustainable Environment: Focus Report on Human Rights and Climate Change. Mapping Report. UNHCR. https://www.ohchr.org/EN/Issues/Environment/SREnvironment/Pages/MappingReport.aspx. Accessed 22 January 2019.

UN Security Council Statement, in Occasion of the 6587th Meeting, 20 July 2011 (doc. no. S/PRST/2011/15).

UNEP et al. (2014). Regional Consultation on the Relationship Between Human Rights Obligations and Environmental Protection, with a Focus on Constitutional Environmental Rights.

UNHCR (2003). Guidelines on International Protection: "Internal Flight or Relocation Alternative" Within the Context of Article 1A(2) of the 1951 Refugee Convention and/or 1957 Protocol Relating to the Status of Refugees. UN Doc HCR/GIP/03/04. https://www.unhcr.org/publications/legal/3f28d5cd4/guidelines-international-protection-4-internal-flight-relocation-alternative.html. Accessed 13 June 2019.

UNHCR (2015). UNHCR, the Environment & Climate Change (Updated Version) (p. 5). Report. UNHCR. https://www.unhcr.org/540854f49.pdf. Accessed 14 June 2019.

UNHCR (2020). UN Human Rights Committee Decision on Climate Change Is a Wake-Up Call, According to UNHCR. https://www.unhcr.org/news/briefing/2020/1/5e2ab8ae4/un-human-rights-committee-decision-climate-change-wake-up-call-according.html. Accessed 10 February 2020.

United Nations University (Institute for Environment and Human Security) (2014). Climate Change and Migration in the Pacific: Links, Attitudes, and Future Scenarios in Nauru, Tuvalu, and Kiribati, Fiji. Report. UN University. https://collections.unu.edu/view/UNU:6515. Accessed 14 June 2019.

United States Government Accountability Office (2009). Alaska Native Villages: Limited Progress Has Been Made on Relocating Villages Threatened by Flooding and Erosion. Report to Congressional Requesters. https://www.gao.gov/assets/300/290468.pdf. Accessed 15 May 2019.

UN-Water (2019). Policy Brief Climate Change and Water. Final Draft. https://www.unwater.org/publications/un-water-policy-brief-on-climate-change-and-water/. Accessed 20 March 2020.

UN-Water, Climate Change Adaptation Is Mainly About Water. https://sustainabledevelopment.un.org/content/documents/UNWclimatechange_EN.pdf. Accessed 13 January 2019.

Van der Vliet (2018). 'Climate Refugees', a Legal Mapping Exercise. In Behrman & Kent (Eds.), *'Climate Refugees' Beyond the Legal Impasse?* New York: Routledge, pp. 23 ff.

Van Wormer & Beshorn (2011). *Human Behavior and the Social Environment: Groups, Communities and Organizations*. Oxford: Oxford University Press, pp. 279 ff.

Vogt (2011). Road to Survival. Cit. in Gemenne, How They Became the Human Face of Climate Change: Research and Policy Interactions in the Birth of the "Environmental Migration" Concept, in Piguet, Pécoud, de Guchtneire (Eds.), *Migration and Climate Change* (p. 227). Cambridge: Cambridge University Press and UNESCO Publishing.

Wang (2017, October 5). Philippines Climate Case Could Find Fossil Fuel Companies Violate Human Rights. *Climate Liability News*. https://www.climateliabilitynews.org/2017/10/05/philippines-climate-change-human-rights/. Accessed 13 May 2019.

Warner et al. (2009). In Search of Shelter: Mapping the Effects of Climate Change on Human Migration and Displacement. Report. Center for International Earth Science Information Network, Earth Institute of Columbia University. http://www.ciesin.columbia.edu/documents/clim-migr-report-june09_media.pdf. Accessed 13 October 2019.

Welch (2015, March 2). Climate Change Helped Spark Syrian War, Study Says. *National Geographic*. https://www.nationalgeographic.com/news/2015/3/150302-syria-war-climate-change-drought/. Accessed 10 June 2019.

Wendle (2015). The Ominous Story of Syria's Climate Refugees. *Scientific American*. https://www.scientificamerican.com/article/ominous-story-of-syria-climate-refugees/. Accessed 14 April 2019.

Westing (1992). Environmental Refugees: A Growing Category of Displaced Persons. *Environmental Conservation*, 19, pp. 201 ff.

Westra (2009). *Environmental Justice and the Rights of Ecological Refugees*. London: Earthscan.

Williams (2008). Turning the Tide: Recognizing Climate Change Refugees in International Law. *Law & Policy*, 4, pp. 515 ff.

Xing-Yin Ni (2015). A Nation Going Under: Legal Protection for "Climate Change Refugees". *Boston College International & Comparative Law Review*, 2, pp. 329 ff.

Zaccal & Lugen (2016). Common but Differentiated Responsibilities Against the Realities of Climate Change. In Papaux & Zurbuchen (Eds.), *Philosophy, Law and Environmental Crisis*. Stuttgart: Nomos.

Zetter (2010). Protecting People Displaced by Climate Change: Some Conceptual Challenges. McAdam (Ed.), *Climate Change and Displacement, Multidisciplinary Perspectives*. Oxford: Hart Publishing, pp. 140 ff.

International Instruments

American Convention on Human Rights, adopted 22 November 1969, entered into force 18 July 1978.

Arab Charter on Human Rights, adopted 15 September 1994.

ASEAN Declaration on Human Rights, adopted 9 November 2012.

Convention Against Torture and Other Cruel, Inhuman or Degrading Treatment or Punishment, adopted 10 December 1984, entered into force 26 June 1987.

Convention on Access to Information, Public Participation in Decision-Making and Access to Justice in Environmental Matters, adopted 25 June 1998, entered into force 30 October.

Convention on Biological Diversity, adopted on 22 May 1992, entered into force 29 December 1993.

Convention Relating to the Status of Refugees, adopted 28 July 1951, entered into force 22 April 1954, to be read in conjunction with the Protocol Relating to the Status of Refugees, adopted on 31 January the 31st 1957.

Draft Principles on Human Rights and the Environment, adopted 16 May 1994, E/CN.4/Sub.2/1994/9, Annex I.

European Convention on Human Rights, adopted 4 November 1950, entered into force 3 September 1953.

Global Compact on Refugees, General Assembly Official Records Seventy-Third Session Supplement No. 12 (A/73/12 [Part II]).

ILC Draft Articles on Responsibility of States for Internationally Wrongful Acts, attached to UN Doc. A/Res/56/83, 12 December 2001.

International Covenant on Civil and Political Rights, adopted 16 December 1966, entered into force 23 March 1976.

International Covenant on Economic, Social and Cultural Rights, adopted on 16 December 1966, entered into force 3 January 1976.

Kyoto Protocol to the United Nations Framework Convention on Climate Change, adopted 10 December 1997, entered into force 16 February 2006, UN Doc. FCCC/CP/1997/L.7/Add.1.

Oslo Principles on Global Climate Change Obligations, released at a symposium at Kings College London on March 30, 2015. https://globaljustice.yale.edu/sites/default/files/files/OsloPrinciples.pdf. Accessed 37 October 2019.

Paris Agreement, adopted 12 December 2015, entered into force 4 November 2016, Decision 1/CP.21, UN Doc. FCCC/CP/2015/10/Add.1.

Peninsula Principles on Climate Displacement within States, Displacement Solutions, 18 August 2013. http://displacementsolutions.org/peninsula-principles/. Accessed 12 February 2020.

The Hague Declaration on the Environment, adopted on 11 March 1989, 28 I.L.M. 1308.

UN Declaration on Environment and Development (Rio Declaration), adopted 12 August 1992.

UN Declaration on Human Environment (Stockholm Declaration), adopted 15 June 1972.

UN Framework Convention on Climate Change, adopted 9 May 1992, entered into force 21 March 1994.

UN General Assembly Resolution (1990), Need to Ensure a Healthy Environment for the Well-Being of Individuals, adopted by the General Assembly on 14 December 1990, UN Doc. A/RES/45/94.

UN General Assembly Resolution (2010), The Human Right to Water and Sanitation, UN Doc. A/RES/64/292.

UN General Assembly Resolution (2015), Sustainable Development Goals: Transforming Our World: The 2030 Agenda for Sustainable Development, UN Doc. A/RES/70/1.

UN General Assembly Resolution (2018), Global Compact for Safe, Orderly and Regular Migration, UN Doc. A/RES/73/195.

UN High Commissioner for Refugees (UNHCR) (1998). Guiding Principles on Internal Displacement. ADM 1.1.PRL. 12.1, PR00/98/109.

UN Human Rights Council Resolution (2011), Res. 16/11, Human Rights and the Environment (UN Doc. A/HCR/RES/16/11).

UN Security Council Resolution 2349 (2017) [on the situation in the Lake Chad Basin region] S/RES/2349.

UN Security Council Resolution 2408 (2018) S/RES/2408 (2018).

Case Law

African Charter's Human Rights and Peoples' Rights Commission (2001), *Social and Economic Rights Action Centre (SERAC) and Another v. Nigeria*, AHRLR 60 (ACHPR 2001).

Corte di Appello di Bologna (2016), 19 February 2016, judgement no. 524/2016.

England and Wales Court of Appeal (2003), *N v. Secretary of State for the Home Department*, case no. C4/2004/0669, 5 August 2004.Civ. 1369, 2003.

European Court of Human Rights (1989), *Söring v. United Kingdom*, Application no. 14038/88, 7 July 1989.

European Court of Human Rights (1998), *Guerra v. Italia*, Application no. 14967/89, 19 February 1998.

European Court of Human Rights (2000), *Labita v. Italy*, Application no. 26772/85, 6 April 2000.

European Court of Human Rights (2002), *Pretty v. United Kingdom*, Application no. 2346/02, 29 April 2002.

European Court of Human Rights (2003), *Hatton v. United Kingdom*, Application no. 36022/97, 10 July 2003.

European Court of Human Rights (2004), *F v. United Kingdom*, Application no. 17341/04, 22 June 2004.

European Court of Human Rights (2004), *Öneryildiz v. Turkey*, Application no. 48939/99, 30 November 2004.

European Court of Human Rights (2005), *Bader and Kanbor v. Sweden*, Application no. 13284/04, 8 November 2005.

European Court of Human Rights (2005), *Fadeyeva v. Russia*, Application no. 55723/00, 30 November 2005.

European Court of Human Rights (2005), *Kyrtatos v. Greece*, Application no. 41666/98, 16 February 2005.

European Court of Human Rights (2006), *Giacomelli v. Italia*, Application no. 59909/00, 19 October 2006.

European Court of Human Rights (2008), *Budayeva et al. v. Russia*, Applications no. 15339/02, 21166/02, 20058/02, 11673/02 and 15343/02, 20 March 2008.

European Court of Human Rights (2008), *Faegerskioeld v. Sweden*, Application no. 37664/04, 26 February 2008.

European Court of Human Rights (2008), *N v. United Kingdom*, Application no. 26565/06, 27 May 2008.

European Court of Human Rights (2011), *MSS v. Belgium & Greece*, Application no. 30696/09, 21 January 2011.

European Court of Human Rights (2011), *Sufi & Elmi v. United Kingdom*, Applications no. 8319/07 and 11449/07, 28 June 2011.

European Court of Human Rights (2012), *Di Sarno et al. v. Italia*, Application no. 30765/08, 10 January 2012.

European Court of Human Rights (2016), *Paposhvili v. Belgium*, Application no. 41738/10, 13 December 2016.

High Court of New Zealand (2013), *Teitiota v. The Chief Executive of the Ministry of Business, Innovation and Employment*, decision of the High Court of New Zealand no. [2013] NZHC 3125 of 26 November 2013.

Inter-American Court of Human Rights (1985), *Yanomani Indians v Brazil*, case No. 7615, 3 March 1985.

Inter-American Court of Human Rights (1999). *Coard et al. v. United States*, case 19.951, Report No. 109/99.

Inter-American Court of Human Rights (2004), *Maya Indigenous Community of the Toledo District v Belize*, case no. 12.053, 12 October 2004.

Inter-American Court of Human Rights (2009), *Kawas-Fernandez v. Honduras*, 3 April 2009.

International Court of Justice (1949), *Corfu Channel Case (UK c. Albania)*, 1949, I.C.J Rep 4, p. 22.

International Court of Justice (1970), *Barcelona Traction* (ICJ, *Barcelona Traction, Light and Power Company, Limited*), I.C.J. Reports, 5 February 1970, p. 3.

International Court of Justice (1996), *Legality of the Threat or Use of Nuclear Weapons, Advisory Opinion*, 1. C.J. Reports 1996, p. 226.

International Court of Justice (1997). *Gabcikovo-Nagymaros case (Hungary v. Slovakia)*, 25 September 2997, I.C.J. Reports 1997.

International Court of Justice (2005). *Armed Activities on the Territory of the Congo (Democratic Republic of the Congo v. Uganda)*, I.C.J. Reports 2005, p. 168.

International Court of Justice, (2004) *Legal Consequences of the Construction of a Wall in the Occupied Palestinian Territory*, Advisory Opinion, I.C.J. Reports 2004, p. 136.

New Zealand Immigration and Protection Tribunal (2013), Decision 800413/2013. https://autlawiel.files.wordpress.com/2014/07/af-kiribati-2013-nzipt-800413.pdf. Accessed February 2018.

New Zealand Immigration and Protection Tribunal (2014), Decision no. [2014] NZIPT 501370-371. https://www.refworld.org/pdfid/585152d14.pdf. Accessed 13 July 2018.

Refugee Review Tribunal of Australia (2009), Decision no. 0907346 [2009] RRTA 1168, 10 December 2009. https://www.refworld.org/pdfid/4b8fdd952.pdf. Accessed 12 June 2017.

The Hague District Court (2015), *Urgenda case* (Rechtbank den Haag, 24 June 2015, C/09/456689 / HA ZA 13-1396). https://uitspraken.rechtspraak.nl/inziendocument?id=ECLI:NL:RBDHA:2015:7196. Accessed 12 May 2019.

UN Human Rights Committee (HRC) (1981), *Sergio Ruben Lopez Burgos v. Uruguay*, Communication no. R.12/15, UN Doc. a/36/40. http://hrlibrary.umn.edu/undocs/session36/12-52.htm. Accessed 13 March 2020.

UN Human Rights Committee (HRC) (2006), *Aalbersberg et al. v. Netherlands*, Communication no. 1440/2005, UN Doc. CCPR/C/87/D/1440/2005. http://hrlibrary.umn.edu/undocs/1440-2005.html. Accessed 10 January 2020.

UN Human Rights Committee (HRC) (2020), *Ioane Teitiota v. New Zealand*, Communication no. 2728/2016, UN Doc CCPR/C/127/D/2728/2016. https://www.refworld.org/cases,HRC,5e26f7134.html. Accessed 7 June 2020.

Index

G. Sciaccaluga, *International Law and the Protection of "Climate Refugees"*,
https://doi.org/10.1007/978-3-030-52402-9

D

F

N

Q

R

S

The manufacturer's authorised representative in the EU is Springer Nature Customer Service Centre GmbH, Europaplatz 3, 69115 Heidelberg, Germany. If you have any concerns regarding our products, please contact ProductSafety@springernature.com

Printed and bound by CPI Group (UK) Ltd, Croydon, CR0 4YY

30/09/2026

02225502-0003